W0091367

Compressors and Modern Process Applications

Compressors and Modern Process Applications

Editor

Aslam Khan

scitus
academics

Compressors and Modern Process Applications

Edited by **Aslam Khan**

Printed in 2017

ISBN: 978-1-68117-344-3

Library of Congress Control Number: 2015939257

© 2016 by
SCITUS Academics LLC,
616, Corporate Way, Suite 2, 4766,
Valley Cottage, NY 10989

www.scitusacademics.com

This book contains information obtained from highly regarded resources. Copyright for individual articles remains with the authors as indicated. All chapters are distributed under the terms of the Creative Commons Attribution License, which permits unrestricted use, distribution, and reproduction in any medium, provided the original author and source are credited.

Notice

Reasonable efforts have been made to publish reliable data and views articulated in the chapters are those of the individual contributors, and not necessarily those of the editors or publishers. Editors or publishers are not responsible for the accuracy of the information in the published chapters or consequences of their use. The publisher believes no responsibility for any damage or grievance to the persons or property arising out of the use of any materials, instructions, methods or thoughts in the book. The editors and the publisher have attempted to trace the copyright holders of all material reproduced in this publication and apologize to copyright holders if permission has not been obtained. If any copyright holder has not been acknowledged, please write to us so we may rectify.

Contents

vi

Preface

Compressors represent a multimillion dollar investment for many plants, and profitability can be neither reached nor sustained by organizations that neglect this critically important asset. This is clearly brought out in more detailed compressor texts; these are available and listed in the references. However, whereas these more detailed texts have often been recommended for and by machinery reliability professionals, a condensed overview of compressor design, operation, and maintenance is desired by other job functions and will thus be given in this book. This material will assist the very wide spectrum of readers whose process involvement brings them into contact with large process compressors. As an example and to run a smooth organization, terminology must be unified and misconceptions dispelled wherever they creep into our thinking.

Editor

Pressure Pulsation Signal Analysis for Centrifugal Compressor Blade Crack Determination

Hongkun Li[1], Xuefeng Zhang[1], Xiaowen Zhang[1], Shuhua Yang[2], and Fujian Xu[1]

[1]School of Mechanical Engineering, Dalian University of Technology, No. 2 Linggong Road, Dalian 116024, China
[2]Shenyang Blower Works Group Corporation, Shenyang 110869, China

ABSTRACT

Blade is a key piece of component for centrifugal compressor. But blade crack could usually occur as blade suffers from the effect of centrifugal forces, gas pressure, friction force, and so on. It could lead to blade failure and centrifugal compressor closing down. Therefore, it is important for blade crack early warning. It is difficult to determine blade crack as the information is weak. In this research, a pressure pulsation (PP) sensor installed in vicinity to the crack area is used to determine blade crack according to blade vibration transfer process analysis. As it cannot show the blade crack information clearly, signal analysis and empirical mode decomposition (EMD) are investigated for feature extraction and early warning. Firstly, signal filter is carried on PP signal around blade passing frequency (BPF) based on working process analysis. Then, envelope analysis is carried on to filter the BPF. In the end, EMD is carried on to determine the characteristic frequency (CF) for blade crack. Dynamic strain sensor is installed on the blade to determine the crack CF. Simulation and experimental investigation are carried on to verify the effectiveness of this method. The results show that this method can be helpful for blade crack classification for centrifugal compressors.

INTRODUCTION

With the development of the society, centrifugal compressor has been widely used in modern industry such as petroleum, chemicals, metallurgy, and aerospace field as an important fluid machine [1]. Meanwhile, centrifugal compressors are developing to be large in scale, high in speed, and automatic in operation [2]. However, the blade failure usually emerges. As the most important part, the impeller transforms kinetic energy into pressure energy. But the impeller suffers from the effect of centrifugal forces, gas pressure, and friction force which usually lead to cracks. According to statistical analysis, 65% centrifugal compressor malfunctions are closely related to the blades. In addition, 40% blade fatigue failures are not fully understood so far [3]. Examples of blade cracks are shown in Figure 1. Fluid-induced vibration is an important factor for blade fatigue failures. It contains acoustic resonance, unsteady flow, rotating stalls, and flutter [4, 5]. Due to the high-velocity flow through the centrifugal compressor and rotating impeller, high-pressure fluctuations occur in the cavity of compressor which could lead the impeller to irregular vibration. Pressure fluctuation acts on the impeller, leading to stress convergence and cracks in the blades. The growing crack will cause blade failure, which results in catastrophes.

Figure 1: Pictures of centrifugal compressor blade cracks.

There are many reasons for cracks on the blades of compressor. Blade cracks are mainly associated with the detection of material, production process, working condition, and high cycle fatigue. So far, researches were mainly concentrated on the defects in the material, processing, and manufacture of impeller causing high fatigue failure. Lourenço investigated the failure of blades [6]. Kermanpur et al. analyzed the failure mechanism of compressor blades made of Ti6Al4V. The results showed fretting fatigue mechanism is the main cause of several premature failures of Ti6Al4V alloyed compressor blades [7]. In recent years, blade cracks caused by excessive alternating stress induced by air-excited vibration have drawn more and more attention from the researchers. In 2007, Eisinger studied the acoustic fatigue which is the coincidence of impeller structural and cavity acoustic modes. The results indicated that the acoustic fatigue would significantly increase the amplitude of vibration and damage the blades [8]. Investigation on alternating stress can be helpful to prevent or reduce the damage from blade cracks [9]. Therefore, condition monitoring and pattern classification are important to prevent blades from failure as well as blade crack detection which could ensure safe operation of the compressor.

It is well known that blade cracks will result in breakdown or even serious accidents of the whole set for centrifugal compressor. It can even lead to heavy losses for a factory. Moreover, personal safety must be considered because the tangential velocity of breakdown blade can be up to 450 m/s. Therefore, incipient classification of blade crack becomes more and more important than ever before.

Traditionally, displacement sensors are introduced to monitor shaft vibration. Meanwhile, vibration-based condition monitoring is also used in shaft crack classification [10, 11]. But it is difficult to recognize shaft cracks only by vibration signals. Moreover, it is impossible to provide any information to characterize blade crack condition from the shaft vibration signals, making blade crack classification more difficult than shaft crack identification. Different methods for blade condition classification have been investigated by many researchers. Liu et al. studied the malfunction identification method of fan blade crack classification by using wavelet packet analysis [12]. Though the structure is similar to centrifugal compressor and fan, centrifugal compressor has good stiffness as the typical difference. Rama Rao and Dutta studied blade crack condition classification for gas turbine blade recognition by using vibration signal information [13]. Yang et al. proposed the auditory spectrum feature extraction using the support vector machine to identify the malfunction of fans [14]. Witek studied the experimental crack propagation for gas turbine blades via vibration signals in laboratory but it was not in a close-loop test-rig [15]. At the same time, some researchers studied wind turbine blade crack classification by using wavelet analysis, scalogram, and so on [16–18]. But it is different from centrifugal compressor blade working condition in speed and load. All these investigations are helpful for blade crack classification, but further study for early warning of centrifugal compressor blade is required. At the same time, air flow experiment is more important for blade condition analysis in real working conditions.

Pressure pulsation (PP) generated by the interference between rotating blades and the stationary vanes contains much information about the blade working conditions and has been used for blade status conditions analysis [19]. However, the crack information in PP signal is weak, so it is difficult to identify patterns just according to time or frequency information, especially for the incipient blade crack condition. Further feature extraction methods are urgently needed for better information collection. Empirical mode decomposition (EMD) is an effective tool for nonstationary signal analysis, which has been widely applied in rolling element bearings and gearbox fault diagnosis. It has great advantage and adaptability in the mechanical fault diagnosis and feature extraction. EMD is a new time-frequency signal analysis method proposed by the scientist of National Aeronautics and Space Administration (NASA) Huang et al. in recent years [20]. This method has been broadly investigated by many researchers since it was provided. It has been applied in different areas for fault diagnosis. Parey et al. used EMD statistical method to detect incipient fault of the gears [21]. Loutridis applied the instantaneous energy density and EMD to monitor and diagnose the gear fault [22]. Liu et al. detected the gear incipient fault with EMD and they found that the result was better than that of wavelet decomposition [23]. EMD can also be used in machine fault diagnosis based on concrete analysis of specific issues. The vibration of blade crack generates a characteristic frequency (CF) which can be modulated into blade passing frequency (BPF). Therefore, EMD can be applied to determine the CF of blade cracks despite of the noise interference in practical working centrifugal compressor. It is also similar to gearbox fault diagnosis problems.

In this paper, PP signals are used for blade working condition classification by using EMD. Experiments are carried on to verify the effectiveness of this method in a test-rig. To verify the effectiveness of this method, strain testing is also carried on for the blade crack analysis. The structure of this paper is as follows. Section 2 introduces the theory of feature extraction for blade crack classification. Section 3 presents the simulation signal analysis. Section 4 describes our experimental setup for blade crack monitoring. Section 5 demonstrates PP signal analysis for blade condition classification. Section 6 gives concluding remarks.

THEORY AND METHOD

Empirical Mode Decomposition

EMD is developed based on instantaneous frequency calculation. It has been considered a very useful tool for the analysis of nonstationary and nonlinear signals [20]. For an arbitrary time series $X(t)$, it can decompose the original into many narrow-band components, each component known as intrinsic mode functions. An intrinsic mode function is used to convert it into a practically useful instantaneous frequency. The intrinsic mode function satisfies two conditions: (1) in the whole range of a data set, the number of the extreme must be equal to the number of the zero crossing points or the difference between them must be one; (2) at any given time, the mean value of the local positive extreme is equal to that of the local negative extreme. An arbitrary nonstationary and nonlinear signal can be decomposed into a series of components satisfied with the intrinsic mode function by using the local wave decomposition method. It is a sifting process and can be written as

$$X(t) - C_1(t) = r_1(t)$$
$$r_1(t) - C_2(t) = r_2(t)$$
$$\vdots$$
$$r_{n-1}(t) - C_n(t) = r_n(t).$$

(1)

The original data can be decomposed into an n-series of intrinsic mode components plus a residual component r_n. The residual can be either a variable or a constant. Thus, the original signal can be expressed as

$$X(t) = \sum_{i=1}^{n} c_i(t) + r_n(t).$$

(2)

After EMD, intrinsic mode function (IMF) can be obtained. FFT can be used on different IMFs analysis for CF determination. EMD can be looked as a filter on feature determination. Therefore, it is helpful to obtain the CF.

Blade Characteristic

In general, centrifugal compressor casing vibration and radiation noise are closely related to blade BPF and its harmonics. It is also generated by the interference between rotor and stator during blade high speed rotation. BPF has high energy in the pressure frequency spectrum. It is the main source of centrifugal compressor noise. Its value can be determined by shaft speed multiplying the number of blade. BPF can be calculated by

$$BPF = \frac{RPM}{60} \times N,$$

(3)

where RPM is the shaft speed and N is the number of blades on the impeller.

BPF is the interference between rotator and stator. As BPF is a high frequency component, the low frequency components for blade nonorder vibration can be modulated to BPF during rotation. The modulation information will appear as the sideband frequency of the BPF. For unbalanced rotor conditions, SF will also be modulated to the BPF giving a sideband frequency around the BPF for unbalanced condition. Sideband frequency could be used to determine the modulated CF. The sideband frequency produced for blade cracks is different from SF. It can be used to warn blade crack. It does not mean that there is a blade with cracks if SF is the sideband frequency for BPF. It is also difficult to classify CF just according to the spectrum for the incipient crack as the magnitude of the blade vibration is weak compared with the amplitude of BPF. Therefore, effective feature extraction is urgently needed for blade crack analysis.

Blade Crack Characteristic Frequency Determination

The steps for CF determination are shown in Figure 2. Firstly, PP is monitored based on the best suitable position according to blade crack classification. This is also a key step to determine the crack information because the sensor location has a direct effect on classification accuracy. Secondly, band-pass filter is applied on signal analysis. Envelope analysis is used to filter BPF signal. Then, IMFs can be obtained by using EMD. Fast Fourier transform is used on different IMFs. In the end, CF for blade crack can be obtained. Blade strain is used to verify the effectiveness of this method.

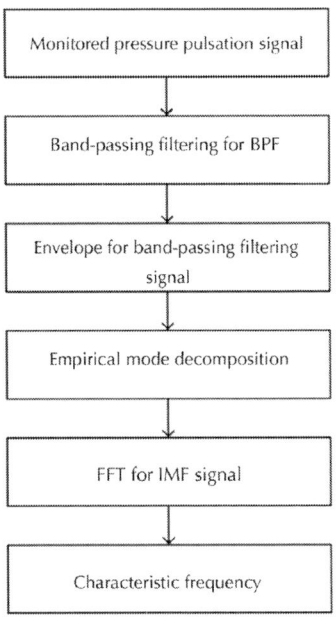

Figure 2: Flowchart for blade crack classification using PP signal.

SIMULATION SIGNAL ANALYSIS

For a amplitude modulation signal sig(t), it can be expressed as

$$\text{sig}(t) = A\left(1 + B\cos\left(2\pi F_e t\right)\right)\sin\left(2\pi F_c t\right),$$

(4)

where, F_c=1500 Hz, F_e=10 Hz, A=60, and B=0.3. F_c and F_e correspond to carrier frequency and modulation frequency, respectively. The corresponding sampling frequency is 10,240 Hz for the simulation signal. Based on (4), an amplitude modulation signal can be obtained as shown in Figure 3(a). Fourier spectrum analysis is shown in Figure

3(b). The main frequency is 1500 Hz. The modulated frequency 10 Hz can be obtained by enlarging the frequency domain around the carrier 1500 Hz frequency shown in Figure 3(c). It is obvious for the sideband frequency around the carrier frequency if there is no noise interference in the signal.

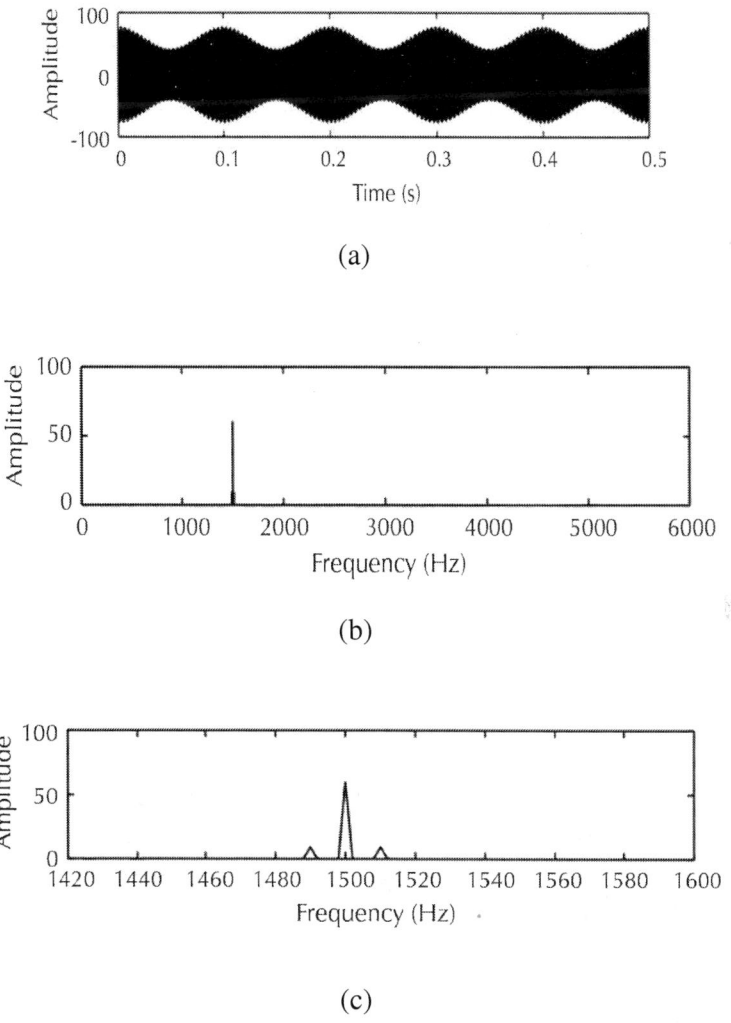

(a)

(b)

(c)

Figure 3: Signal demodulation analysis: (a) time domain signal for the simulation signal; (b) spectrum analysis for the simulation signal; (c) enlarged frequency area for the carrier frequency area.

Strong noise interference is added to the simulation signal as the characteristic information is usually overwhelmed by noise under practical working conditions. The obtained signal is shown in Figure 4(a). In the frequency spectrum analysis, there is clear broad frequency band noise effect shown in Figure 4(b). To determine the modulated signal, the enlargement for carrier frequency area in the spectrum is shown in Figure4(c). Obviously, the enlarged frequency area is not clear due to the noise interference. The noise interference has an effect on the CF determination; therefore, it is difficult to classify the CF just according to sideband frequency spectrum analysis if there is strong noise interference.

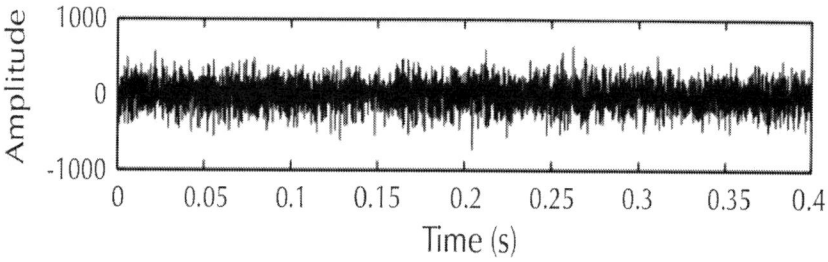

(a)

(b)

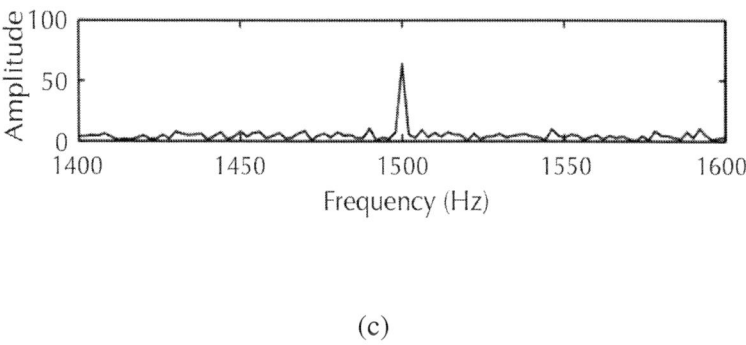

(c)

Figure 4: Frequency spectrum analysis: (a) time domain signal for the noise interference signal; (b) spectrum analysis for the noise interference signal; (c) enlarged frequency area for the carrier frequency area.

Signal filter is used for the monitored signal around BPF. The filter band is 1400–1600 Hz. EMD is applied on the filter signal. IMFs can be obtained as shown in Figure 5. There is not any clear modulated frequency 10 Hz for every IMFs as shown in Figure 6. Envelope method is applied to the filter signal to filter BPF interference. EMD method is also applied on the envelope signal and IMFs can be obtained as shown in Figure 7. But there is clear modulated frequency 10 Hz as shown in Figure 8, the 5th IMFs based on EMD.

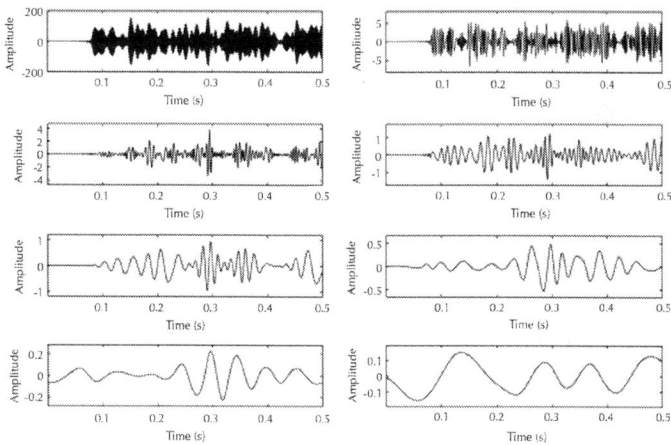

Figure 5: Time domain wave of IMFs based on EMD for the filter signal.

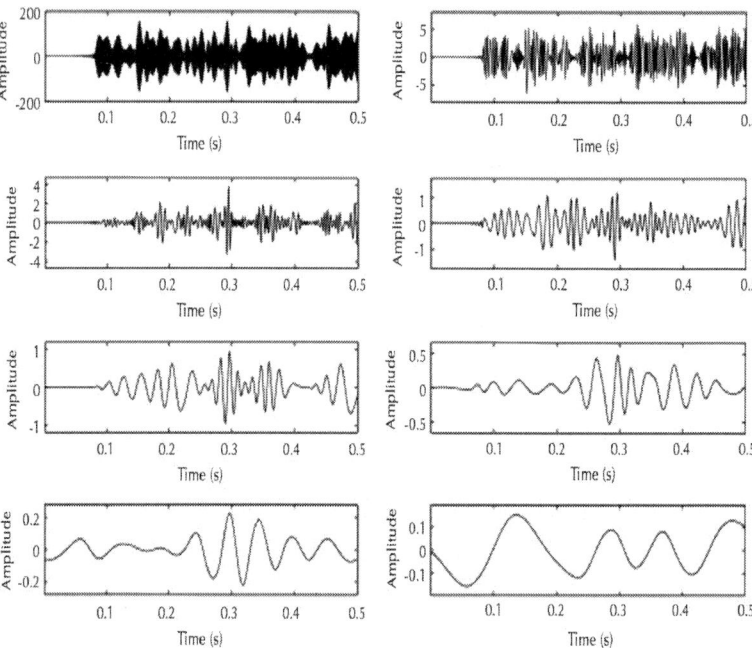

Figure 6: Spectrum of IMFs based on EMD for the filter signal.

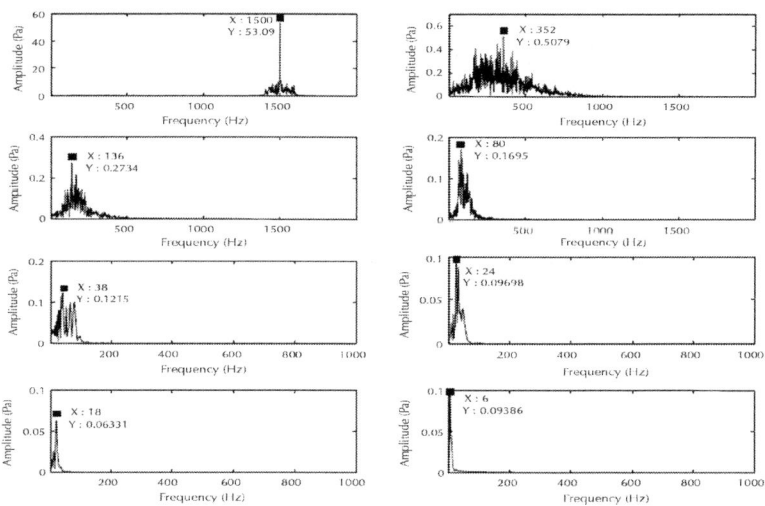

Figure 7: Time domain wave of IMFs based on EMD for the enveloped signal.

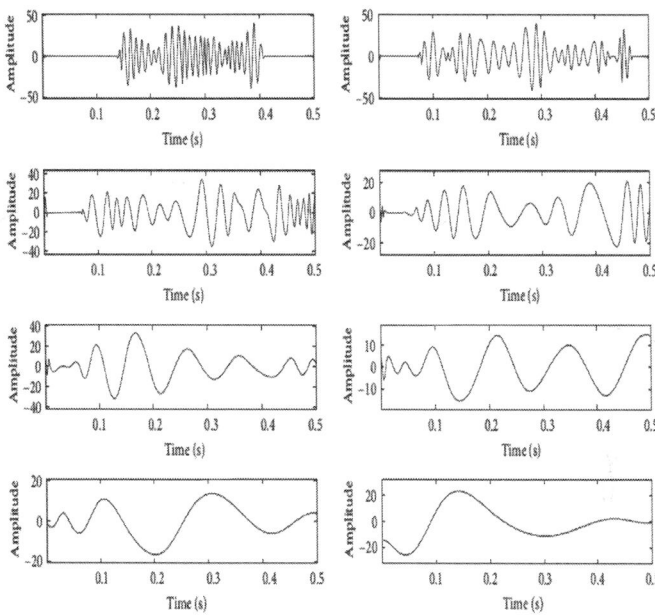

Figure 8: Spectrum of the IMFs for the envelope signal.

Based on above analysis, it can be convenient to determine the modulated frequency though there is strong noise interference. According to above analysis process, the blade nonorder vibration can be also monitored as it has the same property for simulation signal. Therefore, experimental verification is investigated in this research for blade nonorder vibration classification.

EXPERIMENTAL TEST-RIG

Testing-Rig

To verify the effectiveness of this method, an experiment was carried on blade crack condition analysis by using the method based on PP signals analysis in a test-rig. The schematic diagram for the test-rig is shown in Figure 9. It contains an electric motor, fluid coupling, gearbox, and impeller. The impeller is a semiclosed one with 800 mm diameter. It is an experimental impeller for performance testing.

By using fluid coupling, the rotating speed for impeller varies from 500 RPM to 9000 RPM. With the speed-up gearbox, the rotation speed of impeller can meet the designed one. The ratio between the driving and driven gears is 126/43 = 2.93. The experimental picture and hole in the diffuser for installing PP sensor are shown in Figure 10. The crack length during the experiment is 70 mm. PP, vibration, shaft speed sensors are installed to monitor the working process. There are 13 blades in this semiopen impeller. In this experiment, the speed of the impeller is 4500 RPM and 5000 RPM. The SF and BPF parameters are shown in Table 1.

Table 1: Characteristic parameters for φ 800 test-rig

Speed (RPM)	4500	5000
Shaft frequency (Hz)	75	83.3
Blade passing frequency (Hz)	975	1083

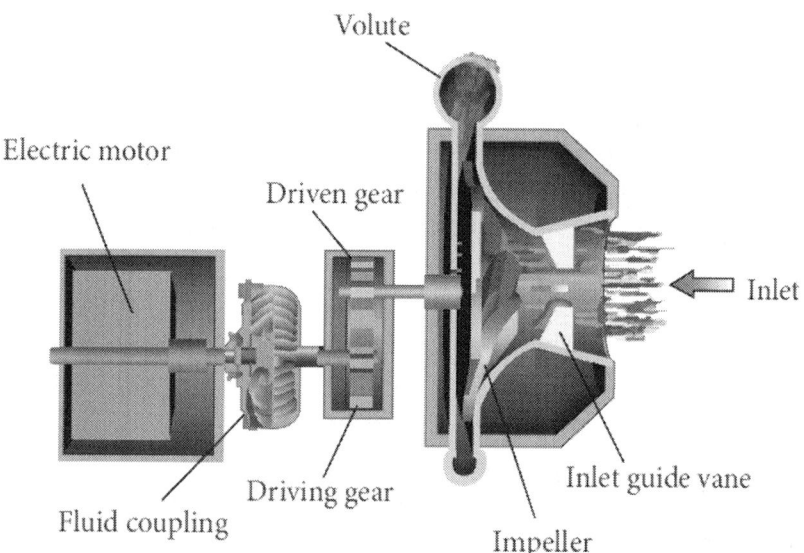

Figure 9: Experimental test-rig.

Figure 10: Pictures for test-rig.

Data Acquisition

There are three PP sensors produced by PCB Piezotronics (New York, USA) to monitor the working process; it is shown in Figure 11. One is installed in the inlet pipe; the other two are installed near the diffuser in the holes shown as Figure 10. The sensitivities of the PP sensors are 0.7044 mV/Pa, 0.9845 mV/Pa, and 0.7336 mV/Pa. PP signal, vibration signal is gathered by the NI-4472 data acquisition card. It is an 8-channel synchronous data gathering system. It is also shown in Figure 11.

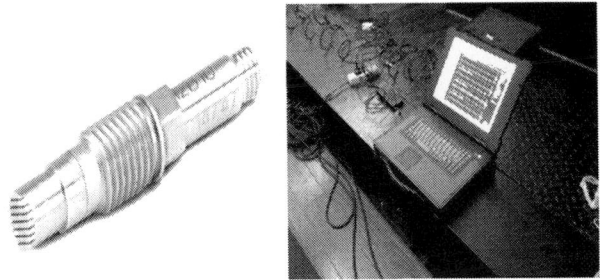

Figure 11: Pressure pulsation data acquisition system.

The blade strain test is carried out by this research to verify the appearance of the fault frequency. The blade vibration will lead to the strain changes on the surface of the blade, so the blade strain can reflect the blade vibration very well. When there are cracks on the blade, the blade vibration due to the change of the blade characteristics such as the stiffness and the blade stress vibration will change at the same time. So the stress test can be used to detect the blade crack. Due to real blade failure process, the location for the crack in selected near the hub shown in Figure 12. So the location for the crack is selected near the hub shown in Figure 12. To verify the appearance of the CF and its relationship with crack, the locations for strain gauge are shown in Figure 12(a). Points (a), (b), and (c) are on the crack blade. Point (d) is on a normal blade. The data acquisition module is shown in Figure 13(a) for launching data. It is also shown in Figure 12(b). Figure 13(b) presents the data receiving module. The data sampling frequency is 1024 Hz for strain signal. There are four channels to monitor the strain shown as Figure 12(a). It is important to determine the blade nonorder vibration as the reason of crack.

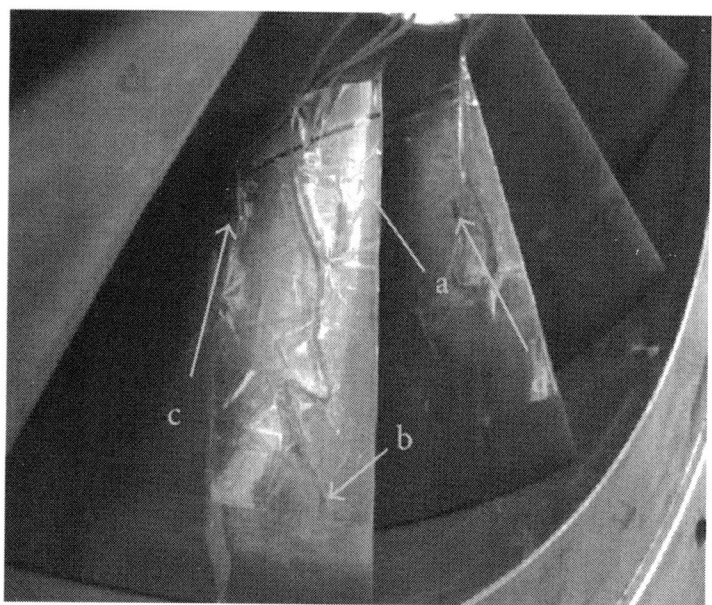

(a)

(b)

Figure 12: Strain testing process.

(a)

(b)

Figure 13: Strain data acquisition system.

DATA ANALYSIS

Strain Signal Analysis

The frequency spectrum for the strain data with impeller speed 4500 RPM is shown in Figure 14. The SF of the impeller is 75 Hz. It is also clear because of the unbalance. The frequency 53 Hz is shown in the spectrum for point (b) and point (c). There is not the CF information for point (d) shown as Figure 14(d) because it is a normal blade. Point (c) is near the crack. It is clearer than point (b). It can be concluded that 53 Hz is the CF for blade vibration.

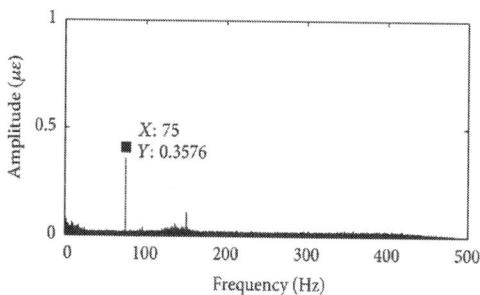

(a)

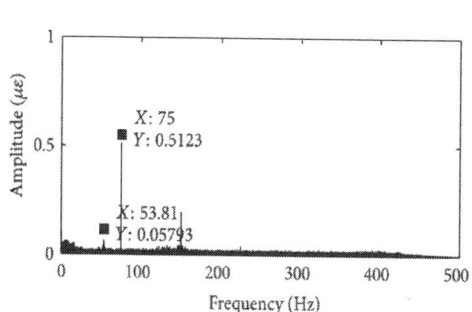

(b)

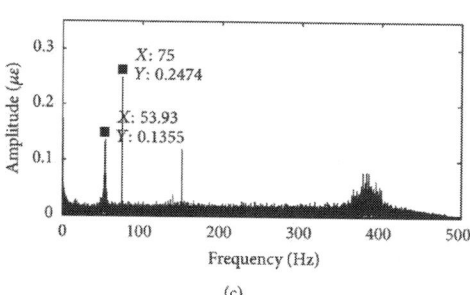

(c)

(c)

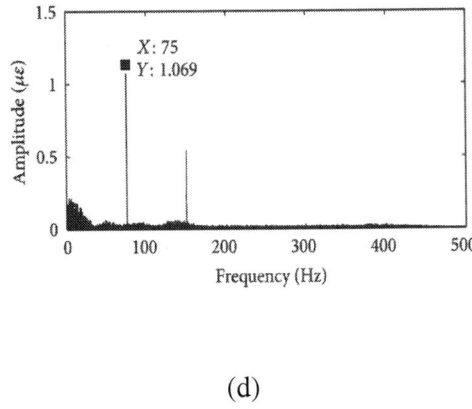

(d)

Figure 14: FFT Strain signal in 4500 RPM.

The same analysis is also carried on the impeller speed in 5000 RPM shown as Figure 15. Based on the above analysis process, the CF for blade nonorder vibration is 52.7 Hz. The CF is almost the same as speed in 4500 RPM. Therefore, it can be concluded that 53 Hz is the CF for the crack. It is a nonorder vibration for the blade and the reason of crack. There is not CF for the normal blade. Strain analysis can help us to determine the CF for blade crack. As it is not convenient in real working condition for strain monitoring, feature extraction is important to obtain the CF from other monitored signals.

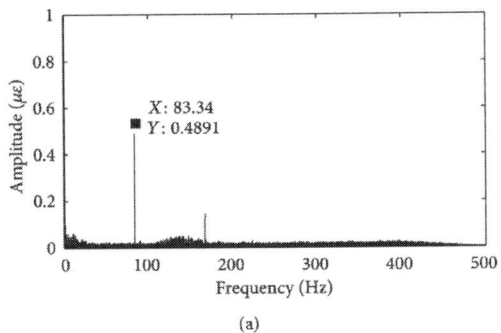

(a)

(a)

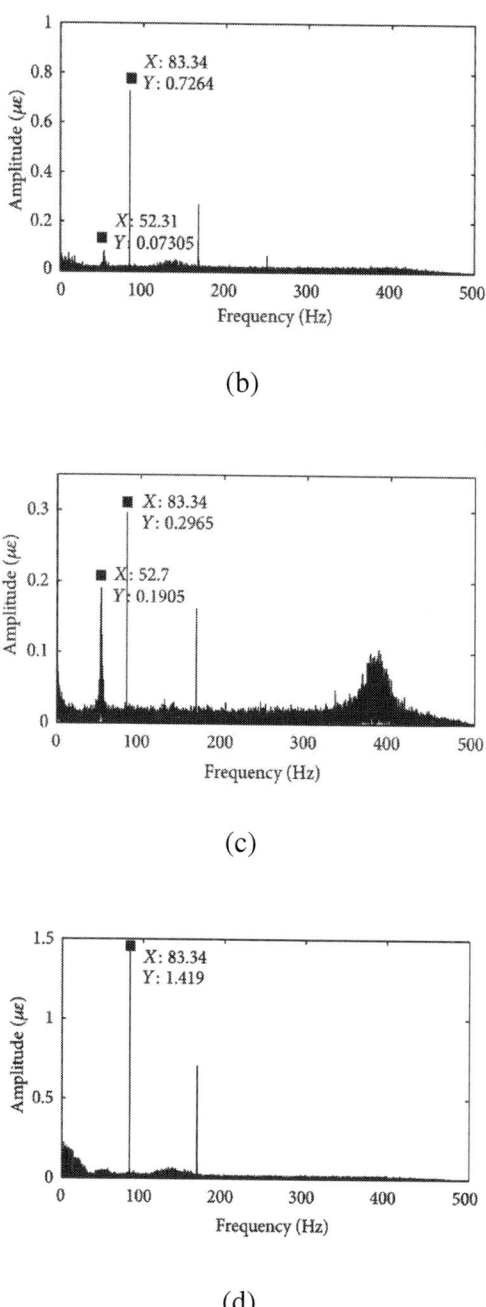

Figure 15: FFT Strain signal in 5000 RPM.

Pressure Pulsation Signal Analysis

PP signal is used to detect blade nonorder vibration information as for the crack. Compared the total length of the blade, the crack is very small (the diameter for the blade is 800 mm) shown in Figure 16. Sides A and B in Figure 16(b) are together with impeller. They are not separated from impeller. It is just for clear demonstration with Figure 16(b). The impeller is manufacturing with whole milling process. At the same time, the averaging thickness of the blade is 10 mm to keep the stiffness of blade. Therefore, the information is weak for blade crack. It is the reason that it is difficult to determine the blade information. It can be just found when there is blade fracture. The crack information will be modulated to BPF as mentioned above although it is weak. There is not any modulated information in time domain shown in Figure 17(a). It is clear for BPF in the spectrum analysis shown in Figure 17(b). But it is not clear for the modulated frequency as the noise interference and nonorder vibration is very weak. It is impossible to obtain the modulated frequency. Therefore, the signal filter is investigated. The filter frequency band is 900–1050 Hz. The filter signal is shown in Figure 17(c). The time domain and the frequency domain spectrum are shown in Figures 18 and 19, respectively. But there is not any information about the CF.

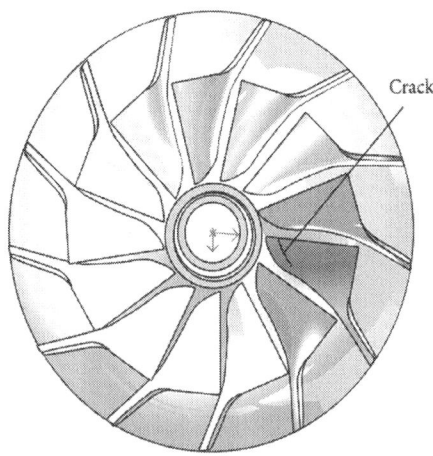

Crack

(a)

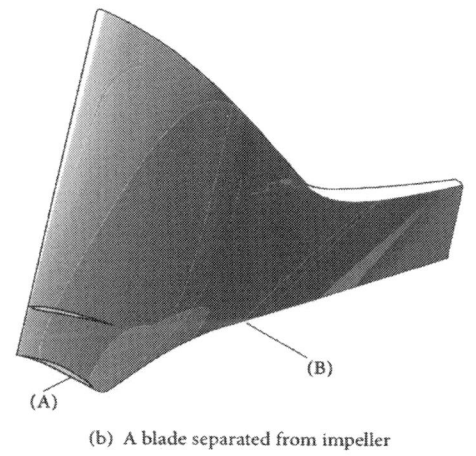

(b) A blade separated from impeller

(b)

Figure 16: Structure of impeller and blade.

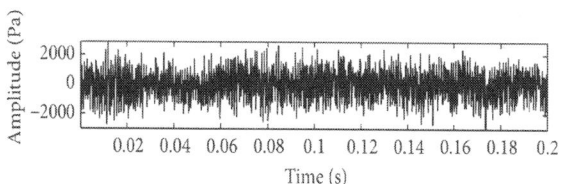

(a)

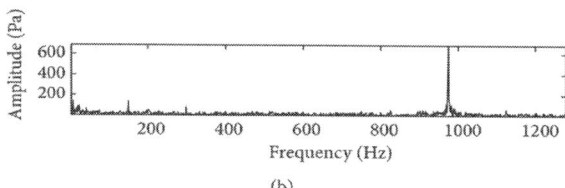

(b)

(b)

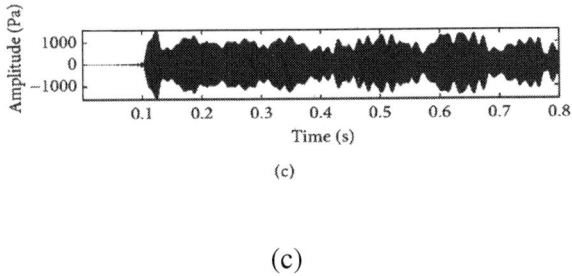

(c)

(c)

Figure 17: Time and frequency domain wave for the PP signal in 4500 RPM.

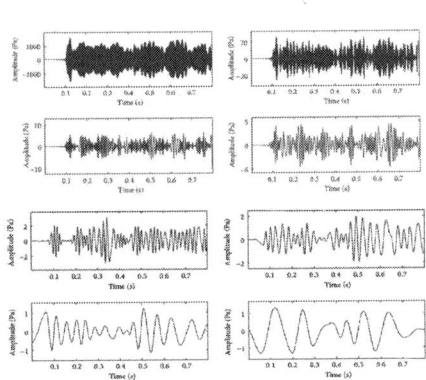

Figure 18: IMFs for the PP signal based on EMD in 4500 RPM.

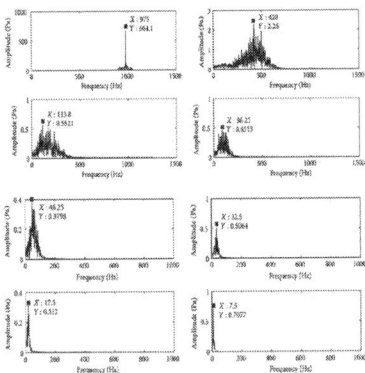

Figure 19: Spectrum for IMFs in 4500 RPM.

Envelope is used on the filtered signal. Then, EMD is used on envelope signal analysis. IMFs can be obtained shown in Figure 20. IMFs frequency spectrum analysis can clearly demonstrate the modulated frequency 53 Hz shown in Figure 21. Therefore, this method can be used to classify the crack CF for blade.

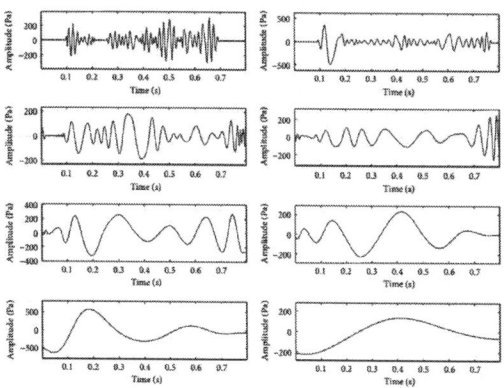

Figure 20: IMFs for the enveloped signal based on EMD in 4500 RPM.

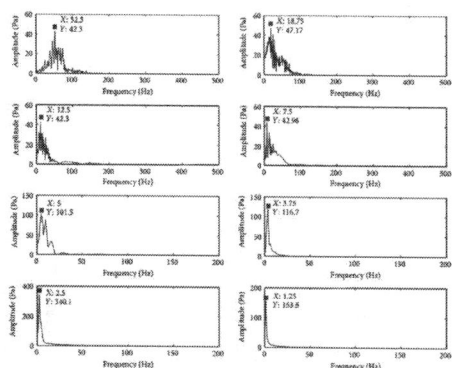

Figure 21: Spectrum for enveloped signal's IMFs in 4500 RPM.

It is also with same result for 5000 RPM. Time and frequency analysis waveform for PP signal is shown in Figure 22. It is also difficult to recognize the CF. Therefore, signal filter is carried on. The filter frequency band is from 990 Hz to 1165 Hz. Then, IMFs based on EMD for envelope signal are shown in Figure 23. IMFs spectrum analysis is shown in Figure 24. It is clear for the modulated frequency 53 Hz. It has the same result with 4500 RPM. It can be verified that this method can effectively recognize the modulated nonorder vibration signal. It also demonstrates that this method can be used on feature frequency determination.

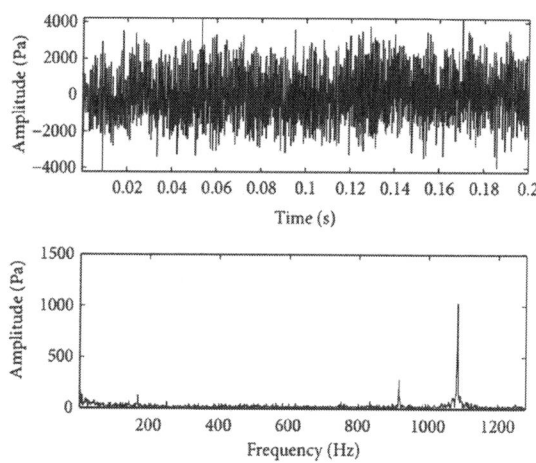

Figure 22: Time and frequency domain wave for the PP signal in 5000 RPM.

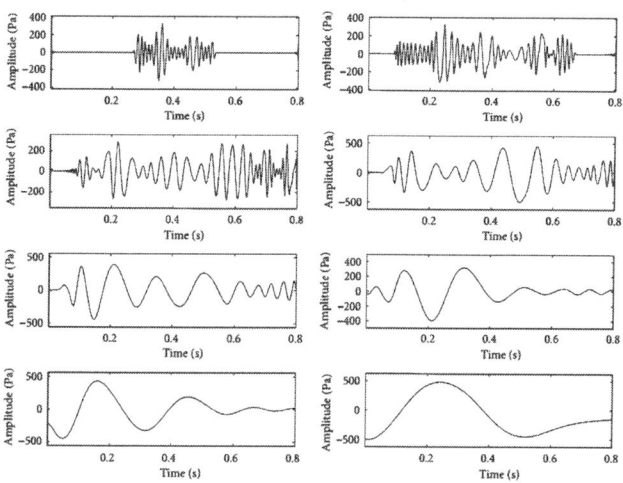

Figure 23: IMFs by EMD for the envelope signal in 5000 RPM.

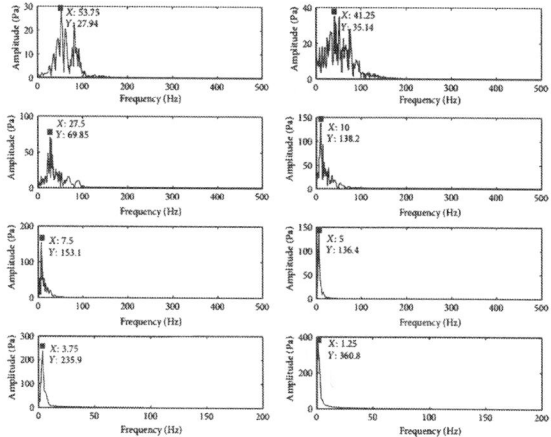

Figure 24: Time and frequency domain wave analysis for the different IMFs in 5000 RPM.

CONCLUSIONS

In this research, PP signals are used for blade crack condition monitoring and classification. The realization of this method is demonstrated in detail. Experiments on an industrial centrifugal compressor with a cracked blade were carried out to verify the effectiveness of this method. CF of blade crack information can be obtained by using EMD and spectrum analysis to obtain the modulated frequency. Strain signal is also investigated to monitor the crack CF. It is verified that crack characteristics can be determined by using PP signal. This research puts forward a method on how to determine the blade crack CF. Further investigations will also be carried on how to apply this method on real working condition blade crack classification. It will be helpful for blade crack early warning.

ACKNOWLEDGMENTS

The work was supported by the Natural Science Foundation of China under Grant no. 51175057 and the Fundamental Research Funds for the Central Universities under Grant no. DUT14ZD204.

REFERENCES

1. Z. Y. Huang and X. F. Wang, Turbine Compressor, Chemical Industry Press, Beijing, China, 2004.

2. A. Kammerer, Experimental Research into Resonant Vibration of Centrifugal Compressor Blades, Swiss Federal Institute of Technology, Zürich, Switzerland, 2009.

3. B. Wen, X. Wu, Q. Ding, et al., Theory and Experiment of Nonlinear Dynamics for Rotating Machinery with Faults, Science Press, Beijing, China, 2004.

4. M. Baumgartner, F. Kameier, and J. Hourmouziadis, Non-Engine Order Blade Vibration in a High Pressure Compressor, ISABE, Melbourne, Australia, 1995.

5. Y. G. Lei, J. Lin, Z. He, and D. Kong, "A method based on multi-sensor data fusion for fault detection of planetary gearboxes," Sensors, vol. 12, no. 2, pp. 2005–2017, 2012. · ·

6. N. J. Lourenço, M. L. A. Graça, L. A. L. Franco, and O. M. M. Silva, "Fatigue failure of a compressor blade," Engineering Failure Analysis, vol. 15, no. 8, pp. 1150–1154, 2008. · ·

7. A. Kermanpur, H. S. Amin, S. Ziaei-Rad, N. Nourbakhshnia, and M. Mosaddeghfar, "Failure analysis of Ti6Al4V gas turbine compressor blades," Engineering Failure Analysis, vol. 15, no. 8, pp. 1052–1064, 2008. · ·

8. F. L. Eisinger and R. E. Sullivan, "Vibration fatigue of centrifugal fan impeller due to Structural-Acoustic coupling and its prevention: a case study," Journal of Pressure Vessel Technology, vol. 129, no. 4, pp. 771–774, 2007. · ·

9. N. Roy and R. Ganguli, "Helicopter rotor blade frequency evolution with damage growth and signal processing," Journal of Sound and Vibration, vol. 283, no. 3–5, pp. 821–851, 2005. · ·

10. K. Elbhbah and J. K. Sinha, "Vibration-based condition monitoring of rotating machines using a machine composite spectrum," Journal of Sound and Vibration, vol. 332, no. 11, pp. 2831–2845, 2013. · ·

11. K. Saravanan and A. S. Sekhar, "Crack detection in a rotor by operational deflection shape and kurtosis using laser vibrometer measurements," Journal of Vibration and Control, vol. 19, no. 8, pp. 1227–1239, 2013. · ·

12. X. B. Liu, J. G. Lin, and Y. Wang, "Research on fault identification of blade crack of fan based on wavelet-packet analysis," Machine Tool & Hydraulics, vol. 35, no. 9, pp. 241–243, 2007.

13. A. Rama Rao and B. K. Dutta, "Vibration analysis for detecting failure of compressor blade," Engineering Failure Analysis, vol. 25, pp. 211–218, 2012. · ·

14. H. H. Yang, H. Hou, X. Y. Zeng, and J. C. Sun, "Fault diagnosis for fan based on auditory spectrum feature of sound signal," Chinese Journal of Scientific Instrument, vol. 30, no. 1, pp. 175–179, 2009. ·

15. L. Witek, "Experimental crack propagation and failure analysis of the first stage compressor blade subjected to vibration," Engineering Failure Analysis, vol. 16, no. 7, pp. 2163–2170, 2009. · ·

16. Y. Qu, C. Z. Chen, X. G. Zhao, and B. Zhou, "Wavelet scalogram identification for crack feature of wind turbine blade," Journal of Shenyang University of Technology, vol. 34, no. 1, pp. 22–47, 2012. ·

17. X. Wang, H. Mao, H. Hu, and Z. Zhang, "Crack localization in hydraulic turbine blades based on kernel independent component analysis and wavelet neural network," International Journal of Computational Intelligence Systems, vol. 6, no. 6, pp. 1116–1124, 2013. · ·

18. B. C. Zhou, C. Zhang, and M. Yu, "Research on dynamic propagating characteristics of wind turbine blade›s cracks," China Mechanical Engineering, vol. 24, no. 8, pp. 1108–1113, 2013. · ·

19. E. Egusquiza, C. Valero, X. Huang, E. Jou, A. Guardo, and C. Rodriguez, "Failure investigation of a large pump-turbine runner," Engineering Failure Analysis, vol. 23, pp. 27–34, 2012. · ·

20. N. E. Huang, Z. Shen, S. R. Long et al., "The empirical mode decomposition and the Hilbert spectrum for nonlinear and non-stationary time series analysis," The Royal Society of London. Proceedings. Series A. Mathematical, Physical and Engineering Sciences, vol. 454, no. 1971, pp. 903–995, 1998. · · ·

21. A. Parey, M. El Badaoui, F. Guillet, and N. Tandon, "Dynamic modelling of spur gear pair and application of empirical mode

decomposition-based statistical analysis for early detection of localized tooth defect," Journal of Sound and Vibration, vol. 294, no. 3, pp. 547–561, 2006. · ·

22.   S. J. Loutridis, "Instantaneous energy density as a feature for gear fault detection," Mechanical Systems and Signal Processing, vol. 20, no. 5, pp. 1239–1253, 2006. · ·

23.   B. Liu, S. Riemenschneider, and Y. Xu, "Gearbox fault diagnosis using empirical mode decomposition and Hilbert spectrum," Mechanical Systems and Signal Processing, vol. 20, no. 3, pp. 718–734, 2006. · ·

The Role of Waste Management in the Control of Hazardous Substances: Lessons Learned

Henning Friege

AWISTA Gesellschaft für Abfallwirtschbvaft und Stadtreinigung mbH, Höherweg 100, D-40233, Düsseldorf, Germany

ABSTRACT

Background

Sorting and disposal of waste are the last steps in the "lifetime" of a product. If products are contaminated with chemicals assessed to be hazardous for man or environment, waste management has the role of a vacuum cleaner in substance chain management working in two different ways: The hazardous compounds have to be properly separated from potential secondary resources in sorting processes. If this is not possible, those products have to be disposed safely. Starting from the experiences collected with some chemicals banned, the tools used for phasing out these chemicals from the technosphere are studied with respect to their influence on the contamination of the environment.

Results

Even if a dangerous substance has been banned, it is further used in a number of products. In the cases presented here, the substances were banned for further use. In the case of CFCs, the substitutes used have partially also been substituted because of adverse effects. Besides the prohibition of use of hazardous substances, numerous other regulations were issued to reduce unsafe handling and minimize emissions into the environment. It turned out that waste management cannot correct mistakes which already happened "upstream" in the product chain. The control of point sources works quite successfully, whereas today the overwhelming emissions stem from diffuse sources, partially caused by unsafe waste management procedures.

Conclusions

Though there are no complete balances for both groups of compounds serving as examples, some conclusions can be drawn based on the experiences collected. Hazardous compounds may be separated successfully from used products or waste,

- If they are mostly used in industry and not in households,
- if they can be identified as part of certain products,
- if their concentration in these products is rather high,
- if technical problems come up when they contaminate secondary raw materials,
- if there is international support for proper waste management.

BACKGROUND

Globalization of consumer products also means global application of chemicals used in these products and also global spread of these chemicals with waste, when the products come to their "end of life". This is not a new experience. In the development of chemical industry, product lines like coal tar dyes were highly significant worldwide just a hundred years ago (see for example[1]). In contrary to natural dyes, tar dyes could be standardized. At the beginning of the 20th century, the German and Swiss chemical industry supplied the whole world with brilliant and reproducible colours based on tar chemistry. Years later, toxic and environmentally hazardous properties of these compounds were discovered. Former areas of tar dyes production are classified as contaminated sites. With the breakthrough of chlorine chemistry, other new chemicals with very interesting properties became commercially successful on a global scale. Again, the detection of hazardous properties and experience with ecotoxicological effects lead to restrictions for the use of important organochlorine substances.

Two fundamental developments have to be mentioned:

The enormous increase of mass flows of chemicals on a global scale.

The global division of labour.

Both developments lead to the more or less global availability also of hazardous chemicals for the production and as ingredients in certain goods and in waste.

In the time lag between the placing of a chemical on the market and regulations after scientific findings about potential hazardous properties, the chemical is spread out in numerous application areas becoming part of the waste after use. How is waste management dealing with hazardous compounds brought into the technosphere? Which are the prepositions for a successful "recall" of a substance of concern? These questions cannot be answered by experiments, but only by the experience collected in case studies. In the following, two prominent groups of substances will be used as examples to answer these questions.

Waste Management as Part of Substance Chain Management

Most commercial products are manufactured using a lot of materials from different origins. Production of goods in a global division of labour is a highly sophisticated management problem. Waste management is mostly solved on a local or regional level. The management of material chains comprises the steps of production, consumption and disposal including reuse, recycling and recovery if possible. Therefore, waste management is only a subsystem in the whole management of material chains [2,3]. What does that mean in daily life? We have to observe some important dilemmas [4,5]:

- A time lag between the placing of the product on the market and the restrictions for use as well as a time lag between the start of production and the disposal of the last products in use.
- The dilution of the hazardous substance as component of products to a greater or lesser extent
- The high dissipation of products with the substance in question in the technosphere.
- The costs for proper disposal of products contaminated by a hazardous component; this is an incentive for mixing it up with normal waste.

Case Study PCBs

Chemical and Physical Properties

Polychlorinated biphenyls (PCBs) are a group of 209 congeners which have been synthesized as technical mixtures (oils, waxes) characterized by their chlorine content (weight %). The general formula is given in Figure 1. The melting points of the technical mixtures are considerably lower than the melting points of isolated congeners. To reduce the viscosity of the higher chlorinated mixtures, often trichlorobenzenes were added. PCBs are very persistent substances, especially higher chlorinated congeners. PCBs tend to accumulate in soil, sediments, and biota, the highest concentrations found in fat of marine predators and human beings. The intake of food, especially animal fat and fish, is the most important contamination pathway for humans.

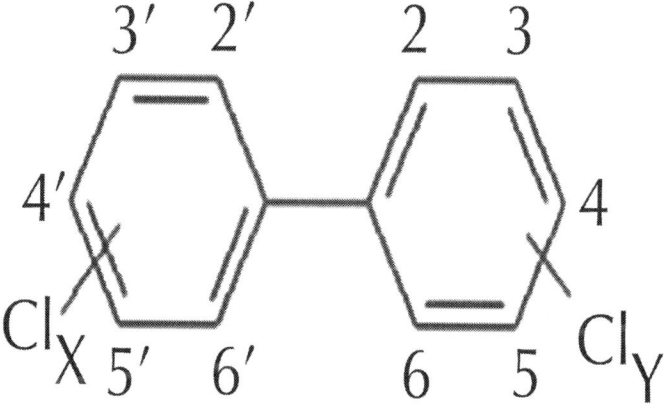

Figure 1: Structure of polychlorinated biphenyls (PCBs).

At a first glance, PCBs seem to be a group of simply structured chemicals. Considering the physical properties, many characteristic features like melting points, partition coefficients and accumulation factors vary depending on the molecular weight and / or the chlorine content. But some properties change considerably with structure – congeners with the same total formula might be very different in toxicity and / or persistence. Some important physicochemical properties are presented in Figure 2:

- The partition coefficients of the congeners differ considerably thus leading to different speed of propagation in the environment.
- The melting and boiling points as well as the K_{ow} increase with increasing substitution of hydrogen by chloride.
- The toxicity of those PCB congeners which have planar structures thus resembling to polychlorinated dibenzodioxins and dibenzofurans is far higher than the toxicity of non-planar PCBs. Therefore, threshold values for food are different for dl-PCBs and ndl-PCBs ("dioxin-like", "not dioxin-like"). Planar structures of PCBs are possible for congeners which have no Cl substitutions in the ortho position to the C-C bond between the aromatic rings.
- The stability of congeners with the same chlorine content varies depending on the substitution patterns: If there are no directly neighboured C-H bonds in a PCB molecule, the stability is quite

higher because the oxidation of carbon atoms in ortho position is not possible.

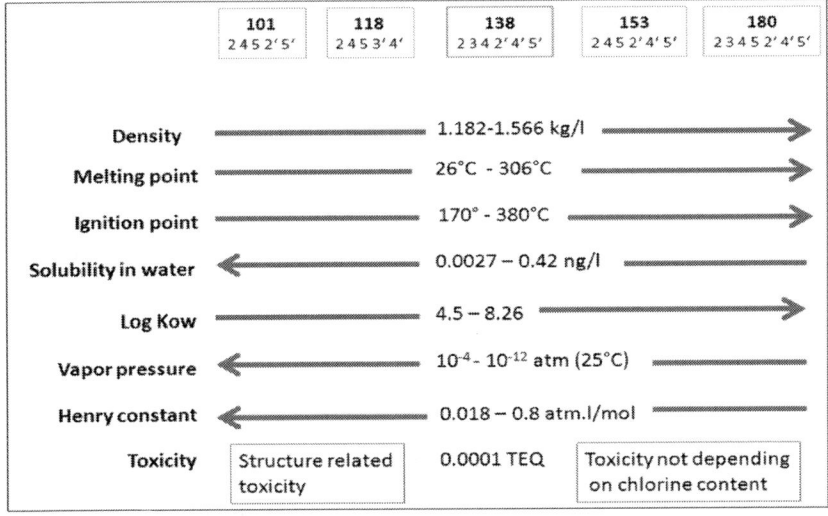

Figure 2: Some important physicochemical properties of PCBs in relation to chlorine substitution (data taken from [6,7]).

Areas of Application and Waste Management

The production of PCBs started in 1930 driven by the need for hardly inflammable isolation material for capacitors and transformers. The most important areas of use are presented in Table1 together with an assessment of the problems coming up with the separation of used products containing PCBs. Many production lines ceased already until the end of the 70ies due to several national PCB bans (Japan), but some producers were busy until the end of the 80ies to meet the demand for special applications or due to delayed legislation. The overall production is estimated at 1.5 Mio Mg. In Europe, PCBs were banned first for "open" [8], then for "closed" applications [9]. But many of the closed applications turned out to be open in reality, as has been demonstrated for hydraulic fluids used in German coal mines, where up to two thirds of the annual consumption of PCBs were lost [10].

Table 1: Former use of PCBs and disposal

	Type of PCB used	Type of Application	Intended way / normal way of disposal	Separation from other items
Isolating agent in transformers	Cl >= 54%	"closed"	Special waste	Easy
Isolating agent in large capacitors	Cl <=42%	"closed"	Special waste	Easy
Hydraulic fluid in mining equipment	Tri- and Tetra-CBs	"closed"	Special waste / waste water from mine draining	Difficult in underground mines
Heat transfer oil in industrial applications	Low chlorinated mixtures	"closed"	Special waste	Easy
Isolating agent in small capacitors	Cl <=42%	"closed"	Waste from electric and electronic equipment / household waste	Possible in state of the art sorting plants
Additive to joint sealer	All technical mixtures	"open"	Hazardous waste / mixed construction waste	Possible in state of the art dismantling processes
Additive to dyes	All technical mixtures	"open"	Hazardous waste / mixed construction waste	Nearly impossible
Additive to textiles	Low chlorinated mixtures	"open"	Hazardous waste / household waste	Nearly impossible
Additive to PVC	High chlorinated mixtures	"open"	Hazardous waste / household waste	Possible in state of the art sorting plants

Friege

Friege *Environmental Sciences Europe* 2012 24:35, doi:10.1186/2190-4715-24-35

Even today, PCBs used for construction are a major source for contamination. About one third of the overall production of technical PCBs was applied for dyes and joint sealers as additives used in millions of buildings worldwide. Normally, details of the construction besides buildings statics are not documented for a long time. Therefore, present owners or inhabitants of older buildings do not know if PCBs have been used in certain materials. Only in the case of secondary contamination of ambient air or food detected by chance or by systematic search, those buildings were decontaminated. There is an unknown number of contaminated areas and buildings left [a].

[a]To give an example: PCB contamination of a building at a German University was reported in 2012: "Verunsicherung an der Uni: Wie gefährlich ist PCB?", WZ Düsseldorf, 15.2.2012.

In contrary to the examples presented above, transformers, heat exchangers and capacitors are "closed systems", if they are not accidentally damaged. As long as machines filled with PCBs had safely closed loops, the use of PCBs was further permitted [9]. This is the case for large electric appliances, which had to be identified by a certain label (e.g. "Clophen", "Arochlor") [11] serving as an information for workers, management, and authorities. This European directive covers all appliances with a volume of more than 5 l liquid including PCBs at concentrations above 50 mg/kg. To protect workers, limit concentrations for workplaces were defined mostly on a national level [12].

From the few examples presented, it is clear that a lot of difficult problems have to be solved mostly by proper waste management:

- Collection and destruction of contaminated building rubble coming from the clean-up of houses.
- Disposal of small capacitors filled with PCB from electric equipment, normally used in discharge lamps and electric household appliances.
- Collection and disposal of large capacitors used for industrial applications.
- Cleaning and sometimes refilling of large transformers.
- Collection and disposal of hydraulic oils completely or partially contaminated with PCBs.

- Collection and disposal of other goods (e.g. PVC parts containing PCB) from households and commerce.

- Reclamation of contaminated sites like former production plants, ruins from large building fires (e.g. department stores, administrative buildings with own electricity supply).

- Last, but not least: under thermal load, chlorine and organic substances react to form PCBs parallel to the "de novo" dioxin synthesis. This happens in particular on the surface of dust particles in the range between 200 and 400°C. A correlation is found between the concentrations of dl-PCBs and PCDDs and PCDFs in the flue gas at the stack [13].

It is obvious that this mission could not be completed successfully. To avoid further contamination of the food chain, limit concentrations were introduced for food [14,15], animal feed, soil, and sludge from waste water treatment [16].

As to the disposal, PCBs can be safely destroyed by incineration in rotary kilns or on the grate if temperatures above 900°C are maintained for some seconds. Legally, waste contaminated up to 50 mg/kg may be incinerated in MWI's. Rotary kilns which are typical for hazardous waste incineration are normally used for disposal of higher PCB concentrations. Former problems with the de novo synthesis of PCBs have been overcome by fast quenching of the flue gas to avoid the temperature window between 200 and 400°C and flue gas cleaning with activated coke. Thus, a modern incineration plant serves as a sink for PCBs and dioxins [17]. There are some other valuable techniques for the destruction of PCBs like the reaction with sodium or potassium, the hydrogenation, or plasma arc treatment. These techniques are fitted only for highly concentrated PCB waste. The disposal in former salt mines is a safe sink for transformers and condensers filled with PCB oils [18,19].

Goods used by millions of people (dissipative use) are hardly collected separately, even if they can be identified by a special form or label. Therefore, nearly all small items containing PCBs have been disposed with household waste. Electric and electronic devices are collected separately since a couple of years to fulfil the WEEE directive. Capacitors containing PCBs and "electrolyte capacitors containing substances of concern" must be separated in sorting plants [20]. These capacitors with a unit weight between 100 and 300 g (a third of this being PCBs) [21] have been used until the end of the 70ies, may be also in the 80ies of the last century. From a technical point of view, more than 95% of all PCB in electronic waste can be separated by the first step in a sorting plant [22]. In a report of the German government, it is assumed that about 3.5 Mg capacitors contaminated with PCBs were separated and disposed in 2008 by proper sorting [23]. But PCB contaminations have been found in sorting plants for e-scrap probably caused by accidentally destroyed capacitors [24,25].

Transformers and large capacitors filled with PCBs can be easily identified by the compulsory label. It is assumed that an overwhelming percentage of these items have been separately disposed. The use of appliances containing PCB with concentrations of more than 50 mg/kg is prohibited since the year 2000; exemptions could be granted until 2010. Instead of disposal of the complete transformer, isolating fluids can be exchanged by other chemicals. The complete cleaning of the contaminated transformer is somewhat difficult and should be done very carefully to avoid increasing PCB concentrations after refilling with other liquids [26]. Due to increasing copper prices, a lot of transformers already disposed underground have been brought back to treatment plants. If the contaminated fluids are completely removed in a cleaning process, this might be a good method to save resources. A German company specialized on this treatment was closed down in 2011 by the responsible authorities because of remarkable pollution of its workers and contamination of the facility [27]. It is likely that there are similar cases in other countries which have either not been uncovered or not documented.

If PCBs are mixed with used mineral oils, the diluted contaminants are spread over a large volume of products. If waste oils are refined for the use as marine diesel, PCBs and its by-products will be emitted by the ships. To avoid this critical pathway, waste oils are also subjected to the European directive 76/769 [8]. Because of the costs for proper disposal of waste contaminated with PCB (actually the price is about 600 €/Mg for large bulks and about 2–3 €/kg for small items [b]), criminal trafficking should be taken into account: Diluting contaminated liquids with other used solvents, faking the waste disposal code, export of contaminated waste for disposal in inappropriate facilities. Actions like this have been severely punished. Transboundary transports of waste contaminated with PCBs are restricted by the Basel convention which established an international regulation for the trading of dangerous waste.

[b]Prices announced by idr-eg, hazardous waste management company, Germany, March 2012.

As to contaminated sites, a lot of experiences have been collected in the last thirty years considering the pathways of different congeners into the environment as well as the metabolism of PCBs. Due to their low water solubility, most congeners do not move from soil to groundwater with the exception of liquid phases together with solvents. If PCBs are spread due to building fires or other accidents, the contamination of workers or residents nearby with PCBs and by-products like dioxins and furans might become a severe problem.

Development of PCB Concentrations in Man and Environment

There is no information about the amount of PCBs still resting in buildings or in other application areas to be disposed in future. We can proceed from the assumption that by far the largest part of PCBs has been disposed or emitted. Following an assessment for Germany, the emissions from the residual appliances were about 221 kg in 2009, whereas nearly 1.7 Mg were emitted in 1990 [28]. This figure does not cover uncontrolled emissions from waste management. After a "clean up"-phase of more than thirty years, the concentrations of PCBs in man and environment have decreased remarkably: As to marine environment, this has been demonstrated recently for gull eggs in German coastal areas covering the period since 1988 [29]. A lot of similar results are obtained by suitable samples documented in the environmental specimen bank. The burden of man is decreasing slowly as is demonstrated in Table 2 for blood of young people (students from four universities). On the basis of an enormous number of milk samples, Fürst [30] demonstrated a continuous decrease of PCB concentrations after a peak between 1985 and 1990 and a second increase of concentrations at the end of the 90ies. The concentrations of the persistent highly chlorinated congeners decrease slowly in comparison with lower chlorinated congeners. But the ingestion of PCBs by food in Germany, especially fish, milk, eggs, meat, and vegetables is still in the range of the tolerable weekly intake [31].

Table 2: Concentration of important PCB congeners in blood plasma of German students (ng ml^{-1} fresh weight); data from the German environmental specimen banking [32]

Sample origin	1997			2010		
	PCB 138	PCB 153	PCB 180	PCB 138	PCB 153	PCB 180
Münster	0.91	0.55	0.43	0.19	0.21	0.14
Greifswald	0.68	0.52	0.31	0.15	0.16	0.11
Halle	0.74	0.44	0.30	0.15	0.17	0.11
Ulm	0.62	0.45	0.34	0.15	0.17	0.12

Friege

Friege *Environmental Sciences Europe* 2012 24:35, doi:10.1186/2190-4715-24-35

Case Study CFCs / HCFCs

Physical and Chemical Properties

Freon is a common name (trade mark by Dupont) for chlorofluorocarbons (CFCs) the molecular structure mostly consisting of one or two carbon atoms. At room temperature, most of these are in gaseous, some in liquid state of matter. The boiling points increase with increasing chlorine content of the molecule. CFCs are non-flammable. All halocarbons of this type are fugitive and also very persistent with half-life periods in the atmosphere of >> 10years. Due to scission of the molecules caused by UV rays in the stratosphere, Cl$^{\circ}$ radicals are produced which react with ozone (chain reaction) thus decreasing the ozone concentration considerably (ozone depleting substances – ODS). Some are also very powerful greenhouse gases with GWP factors up to 10,000 as compared to CO_2. Hydrochlorofluorocarbons (HCFCs) have similar properties, but are less dangerous with respect to ozone depletion. These substances have no relevant toxic properties with respect to man or animal. Many HCFCs are also not-flammable. Some prominent members of this family are presented in Figure 3.

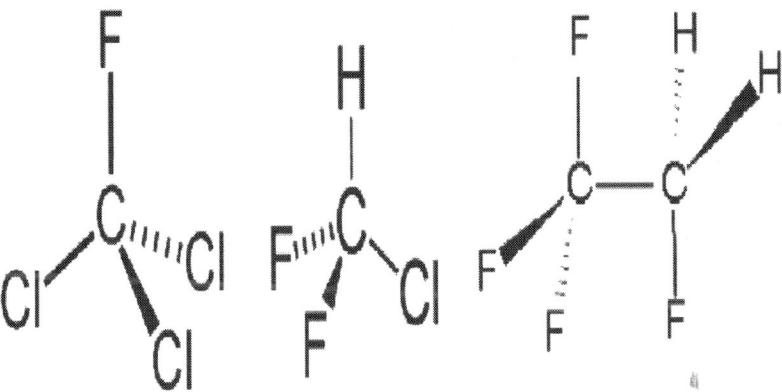

Figure 3: Trichlorofluoromethane (F11), Chlorodifluoromethane (F22), and 1,1,1,2-Tetrafluoroethane (R134a).

The reactions of CFCs in the stratosphere leading to an increase of UV rays became an issue in the public, probably because of the popular picture of the ozone hole above the South Pole. Increasing UV ray intensity means also a higher probability for skin cancer, but epidemiological data do not provide a correlation for Australian people [33]. The ozone depletion potential (ODP) and the greenhouse warming potential (GWP) and the global emissions of several important members of the Freon family may be taken from Table 3.

Table 3: ODP (in relation to F-11) and GWP (in relation to CO_2) for some typical Freon compounds (data from [34], ODP/GWP of F-22 [35])

	Formula	Global emissions in 2008 [Gg yr-1]	ODP	GWP
F-11	CCl3F	52-91	1.0	4,000
F-12	CCl2F2	41-99	1.0	8,500
F-22	CHClF2	385-481 (sum of all HCFCs)	0,05	1,700
F-23	CHF3	12	0	14,200
F-134a	CF3 -CH2F	149±27	0	1,370

Friege

Friege *Environmental Sciences Europe* 2012 24:35, doi:10.1186/2190-4715-24-35

Areas of Application and Waste Management

By expanding Freon chemicals, they cool down (Joule Thomson effect). Freon substances are therefore excellent cooling agents; in the mid of the 20th century, they substituted toxic ammonia and inflammable methyl chloride as common refrigerants for freezers. Due to their good solubility for many organic compounds, Freon substances are used in chemical reactions under pressure, i.e. for extractions. The use of Freon gases as foaming agents for polystyrene or polyurethane improves the isolating properties of the polymer foams. Freon compounds have been banned in international conventions ("Montreal protocol" [36]) starting in 1987. The Montreal Protocol signed in 1987 has been amended several times to accelerate the phase-out of certain substances, to include additional compounds, or to change control mechanisms. It is the first international treaty to achieve universal ratification by all 196 UN members. The signatory countries are obliged to reduce the production of Freon substances stepwise in relation to former production volumes. The use of individual compounds is phased out within reasonable time frames. The countries have to report production, import, export and destruction volumes to the Secretariat of the Protocol. In Table 4 the most important use pattern of fluorinated hydrocarbons is presented. By contrast to PCBs, CFCs and especially HCFCs are further produced in some countries. The substitution process is continuously going on by phasing out CFCs partially replaced by HCFCs thus increasing the banks of powerful greenhouse gases. Therefore, the substitution of CFCs by substances with low ODP as well as low GWP is of outstanding importance.

Table 4: Use pattern of CFCs, HCFCs and fluorinated hydrocarbons

Area of use	Type of chemical used	Application
Cooling agent (freezers, refrigerators...)	F-11 → HCFCs → Pentane	Closed

Propellant gases for aerosols	F-11, F-12 → F-134a and other fluorinated hydrocarbons → Propane/ Butane → pressurized air	Open
Foam blowing agent for polyurethane and polystyrene	F-11, F-12 → F-22 → F-134a, F-152a	Open
Air condition for automobiles	F-22 → F-134a → 2,3,3,3- → CO2	partially open

Friege

Friege *Environmental Sciences Europe* 2012 24:35, doi:10.1186/2190-4715-24-35

The Montreal protocol and its follower conventions focus on a production ban with controlled exceptions (e.g. from security reasons or laboratory use) of chlorinated and fluorinated hydrocarbons in relation to their ODP. The parties to the protocol including the European Community are not directly obliged to collect Freon from used products, but the governments have "to take appropriate decisions, consistent with the objectives of the Basel Convention and of the Montreal Protocol, in order to facilitate early phase-out of the production and consumption of the controlled substances of the Montreal Protocol" [37]. Thus, the Montreal Protocol accepts CFC emissions from banks. To decrease emissions from leakages, the German government enacted a regulation for large cooling and air condition systems to ensure regular maintenance by skilled personal [38].

Freon gases from refrigerators, freezers, air condition equipment, and heat pumps shall be recovered for destruction, recycling or reclamation [39]. The destruction of Freon is regulated to avoid improper disposal techniques: Rotary kiln is approved for all types of ODS, municipal waste incineration is approved only for foams, and some plasma arc methods are approved for ODS with the exception of foams [39]. With respect to the WEEE directive [20], used refrigerators, freezers, and other electric and electronic devices have to be collected separately. The producers are obliged to take back used items and to pay for recycling and disposal. German authorities [40] urge companies running e-waste recycling facilities to recover 90% of the Freon compounds from used devices including the isolation foam of freezers. To prevent improper recycling, the export of unusable devices in non-EU countries has been prohibited. In daily life, the success of these efforts is rather poor: The yearly minimum collection target within the EU is 4 kg per inhabitant (4 kg $inh^{-1}yr^{-1}$) with respect to all waste from electric and electronic equipment. In 2008, about 6.95 kg $inh^{-1}yr^{-1}$ have been collected in the European Union, whereas the overall amount of waste from electric and electronic devices was about 20.8 kg inh^{-1} yr^{-1}. (The latter figure has been calculated using an empirical correlation published by Huisman [41].) According to European statistics, freezers and refrigerators are recorded in the category of large household appliances ("white goods"). Therefore, we have only few data concerning the recycling of used appliances containing Freon compounds. Actually, most of the devices to be disposed contain either F-11/F-12 or the successor products, i.e. partially halogenated hydrocarbons like F-22 or fluorinated hydrocarbons. In new refrigerators, pentane is used as cooling agent. In the State of Nordrhein-Westfalen (17.8 Mio inhabitants), 550,000 to 600,000 refrigerators have to be disposed per year. Until 2014, more than 50% of these devices will be contaminated with CFCs because of the long lifetime of household refrigerators [42]. Due to compulsory Freon elimination before metal recovery, the costs for the recycling of freezers are normally higher than the earnings from scrap. Therefore, refrigerators which do not work any longer (broken) cannot be sold by their owners. This might be the reason for the relatively high amount of refrigerators collected in comparison with other household appliances reported from the Dutch waste collection [43]. A considerable percentage of refrigerators and freezers are severely damaged by informal collectors who cut out copper tubes and compressors. Statistical data about these emissions are normally not found in government reports. Following an assumption by a German NGO, about 45% of Freon compounds from household appliances are emitted [44]. The problems show primarily up with kerbside collection. In the city of Düsseldorf (~ 600,000 inhabitants), about 70% of the freezers and refrigerators declared for e-waste collection are damaged or stolen [45]. But also appliances brought to recycling yards are sometimes damaged. In the recycling facilities, Freon compounds used in the compressor and in the isolation must be separated in a first step. The recovery target for these contaminants is

90% according to German authorities [40]. Following some random samples [46], the recovery targets are not met by most recycling facilities. The emissions from German point sources in 2009 add up to 0.82 Mg CFCs and 21.9 Mg HCFCs, respectively (data taken from [47]). Assuming a mean amount of 250 g Freon in a refrigerator, 2.5 million items to be disposed per year of which 50% are filled with CFCs and 45% loss during collection, the overall emission from this source is about 140 Mg per year. Thus, the diffuse sources mainly caused by improper waste management are far more important than the point sources.

There are no comparable regulations for construction and demolition waste, though the amount of CFCs used in polymer foams is also important.

Another interesting example is the case of R-134a (see Tables 3 and 4) which is used since the nineties in automotive air condition appliances as a substitute for F-22. Due to lack of tightness of these appliances, part of the cooling agent used is emitted. According to a German study based on reports of car repair shops the losses added up to 8.2% per year with respect to the year 2000[48].

Development of Environmental Concentrations of CFCs and HCFCs

The amount of chemicals with high ODP produced worldwide has decreased following the Montreal protocol and further international treaties. Because of the high volatility of the substances in question, the real decontamination of the technosphere can only be demonstrated by measurements in the troposphere followed by model calculations for the development of the ozone concentration. As a result of the implementation of the conventions mentioned before, the concentration of chlorine and of fully halogenated CFCs is decreasing, whereas the amount of partially halogenated chemicals like HCFCs is increasing again. The influence of improper waste management on the overall development cannot be assessed quantitatively. But closing the sources of illegal emission of Freon compounds would help to revitalize the earth's ozone belt years earlier. The decontamination of the stratosphere will need a lot of time. Actually, the ozone layer is expected to return to its pre-1980 levels around the middle of the 21st century [49].

DISCUSSION

The ban of hazardous substances or severe restrictions for their use is the most important prerequisite for the decontamination of the environment. As has been also demonstrated in the examples above, the cleanup phase is delayed.

- by the lifetime of products containing the contaminants in question,
- by the amount of these chemicals emitted during their life cycle,
- by the persistence of the emitted substances in natural environment.

When the used products leave households, workshops, and industrial facilities, identification and separation from other waste is often very difficult. In the cases reported here, "open applications" of these substances being critical to man and/or environment were banned in a first step to avoid further emission into the environment (e.g. use in aerosols, paintings, foams, liners). In both cases reported here, the predominant part of contaminants was used in "closed applications". As may be taken from the use of PCBs in mining equipment and the use of R-134a in automotive air conditioning, many applications in so-called closed systems turned out to be open in reality. Therefore, the use of substances in closed systems has to be controlled carefully to avoid unwanted emissions.

In the case of PCBs, numerous regulations have been issued to avoid the mixture of waste containing PCBs with other waste. This worked from several reasons:

- As to equipment filled with PCBs for further use, the PCB content was indicated by a label (trade name "Clophen", "Arochlor"). In combination with the limit values established for workers' protection, the correct disposal of these items could be assured.
- If the limit of PCBs in used oils is exceeded, the owner is enforced to destruct the complete oil batch by incineration in rotary kilns. Acting against the law is severely punished. Therefore, mixing up of low contaminated oil with high contaminated oil does not pay and is looked at as a crime.
- Contamination of poultry after application of polluted fodder leads to destruction of food, if the limit values for food are

exceeded. Farmers became therefore cautious with regard to their animal feed.

- Transboundary shipment of PCB waste is regulated by the Basel convention.

As demonstrated for e-scrap, PCBs can only partially be separated due to accidental damage of appliances and lack of information about the use of capacitors filled with PCBs in household equipment.

In the case of Freon compounds used in closed applications like compressors, the regulations for safe disposal do not work successfully mostly due to activities of the informal sector. In contrary to large appliances for commercial use (transformers,…), the enforcement of regulations for highly dissipated devices is very difficult. Therefore, the informal sector acts more or less uncontrolled by the responsible authorities [c]. As to "open" applications of Freon substances like propellant gases, only the residual content from spray cans is destructed using appropriate disposal techniques.

[c]From the author's experience, many authorities do not officially report these problems, because they are not in a position or not willing to prosecute environmental crimes by the informal sector.

There is another interesting parallel experience collected with both groups of chemicals: Especially in the case of electric appliances, these chemicals contaminate precious metals, mostly copper. Recent local contaminations by improper recycling of transformers which had been safely disposed years before [27] have been propagated by the intrinsic value of metals in used devices like transformers. Increasing prices for metals on the market will therefore enhance unsafe waste management practices.

It is necessary to establish safe procedures for the disposal of contaminated waste. In both cases presented, these procedures have been developed. Besides some chemical destruction methods and plasma arcs rarely used, incineration $> 900°C$ or $> 1200°C$ with recovery of the halogens (HCl, HF,…) is state of the art.

CONCLUSIONS

When the "life cycle" of a product comes to its end, waste management is responsible for proper handling, recycling of valuables in the waste, and disposal of hazardous compounds in the product. Very often, a separation of valuable and hazardous compounds in those products is not possible. Substances banned for further use will show up in the waste chain even if the products are out of the market since a long time. The time tables for phasing out these substances are often very different depending on national legislation. Moreover, people in industrialized countries, especially skilled workers, are more aware of harmful substances in products as people in developing countries. Globalization of trade without harmonizing the rules for the use of hazardous substances may therefore end up with serious contamination of secondary raw materials by hazardous substances which have not been properly separated in the waste chain.

Hazardous compounds may be separated successfully from used products or waste,

- If they are mostly used in industry and not in households,
- if they can be identified as part of certain products,
- if their concentration in these products is rather high,
- if technical problems come up when they contaminate secondary raw materials,
- if there is international support for proper waste management.

If separation of hazardous materials and resources is not possible, state of the art incineration (WtE) is the approved method for the disposal of hazardous organic waste as well as of contaminated products.

Lack of information about potential contamination of used products may be one reason for unsafe procedures in waste management. It turned out that regulations for workers' protection against the substances in question facilitate the identification of contaminated areas and equipment. Therefore, capacity building and right to information for workers seems to be an important tool not only for the protection of man, but also for better waste management at the origin.

Material input for long-range application should be very carefully documented to facilitate safe deconstruction. The largest material streams result from pulling down of houses or infrastructure. Contamination of debris can be avoided if deconstruction based on detailed information about the construction materials is carried out. Therefore the introduction of "construction passports" would be very helpful for the documentation of materials used for construction and refurbishment. The construction passport would close the knowledge gap between the start and the end of the products' lifetime.

Activities of the informal sector may considerably disturb the "cleanup" of the technosphere by collection and recycling of products contaminated with hazardous substances. There is no unique recipe how to deal with those activities: disseminating knowledge among people in the informal sector might be one important step. But the responsible authorities should also enforce separate collection of hazardous used products and hazardous waste by skilled personal to avoid unsafe recycling procedures.

The new European "waste hierarchy" is a general management rule for the waste sector. From the results presented here, it is concluded that waste management of products should be supported by mass flow balances for hazardous chemicals as well as for scarce non-renewable resources. The equations "recovery is better than disposal" and "material recovery is better than energy recovery" may lead to problems for man and environment if used without further insight into the life cycle of the products in question.

The question arises, if problems like those presented here may show up again despite the progressive chemical policy. Due to the principles followed by REACH, the responsibility of the producers and importers for the safe use of their products has been enhanced. Manufacturers introducing a new application field for an already registered compound are obliged to re-register. The experiments to be executed before registering and marketing of new chemicals should minimize the risk of severe misjudgements about the hazards also in case of special appliances. But errors cannot be excluded. As to the prominent case of nano materials, the registration is under discussion, because they might behave in another way than the molecules in macroscopic structure.

REFERENCES

1.   Henseling KO, Salinger A: Teerfarben - Keimzelle der modernen Chemieindustrie. In Das blaue Wunder. Zur Geschichte der synthetischen Farben. Edited by Andersen A, Spelsberg G. Köln: Kölner Volksblatt Verlag; 1990.

2. Brunner PH, Rechberger H: Practical Handbook of Material Flow Analysis.: Boca Raton; 2004. ISBN 1-5667-0604-1.

3. Friege H: Von der Abfallwirtschaft zum Management von Stoffströmen. Müll und Abfall 1997, 29(1):4–13.

4. Friege H: Ressourcenschonung am Beispiel der Elektro- und Elektronikaltgeräte I. Grenzen des WEEE-Ansatzes. Müll und Abfall 2012, 44(2):80–93.

5. Friege H: Ressourcenschonung am Beispiel der Elektro- und Elektronikaltgeräte II. Ansätze für einen effizienteren Umgang mit nicht erneuerbaren Ressourcen. Müll und Abfall 2012, 44(6):307–317.

6. Fiedler H: Polychlorinated Biphenyls (PCBs): Uses and Environmental Releases: UNEP; 1997. http://www.chem.unep. ch/pops/pops_inc/proceedings/ bangkok/fiedler1.html, accessed 12.2.2012.

7. Li N, Wania F, Lei YD, Daly GL: A Comprehensive and Critical Compilation, Evaluation, and Selection of Physical-Chemical Properties Data for Selected PCB. J Phys Chem Ref Data 2003, 32(4):1545–1590.

8. EC: Council Directive 76/769/EEC of 27 July 1976 on the approximation of the laws, regulations and administrative provisions of the Member States relating to restrictions on the marketing and use of certain dangerous substances and preparations. OJ L 1976, 262:201–203.

9. EC: Council Directive 85/467/EEC of 1 October 1985 amending for the sixth time (PCBs/PCTs) Directive 76/769/EEC on the approximation of the laws, regulations and administrative provisions of the Member States relating to restrictions on the marketing and use of certain dangerous substances and preparations. OJ L 1985, 269:56–58.

10. Rauhut A, quoted after Friege H, Nagel R: Umweltgift PCB. Freiburg: BUND-Information 21; 1982.

11. EC: Council Directive 96/59/EC of 16 September 1996 on the disposal of polychlorinated biphenyls and polychlorinated terphenyls (PCB/PCT). OJ L 1996, 243:31–35.

12. Technische Regeln für Gefahrstoffe. Arbeitsplatzgrenzwerte (TRGS 900). http://www.baua.de/de/Themen-von-A-Z/Gefahrstoffe/ TRGS/pdf/TRGS-900. pdf?__blob=publicationFile&v=8.

13. Johnke B, Menke D, Börske J: Neue Bewertung bei den Toxizitätsäquivalenten für Dioxine/Furane und für PCB durch die WHO. UWSF – Z. Umweltchem. Ökotox 2001, 13(3):175–180.

14. EC: Commission Regulation (EC) No 1881/2006 of 19 December 2006 setting maximum levels for certain contaminants in foodstuffs. OJ L 2006, 364:5–24.

15. EC: Commission Regulation (EC) No 565/2008 of 18 June 2008 amending Regulation (EC) No 1881/2006 setting maximum levels for certain contaminants in foodstuffs as regards the establishment of a maximum level for dioxins and PCBs in fish liver. OJ L 2008, 160:20–21.

16. Klärschlammverordnung, zuletzt geändert am 9.11.2010. http://www.   gesetze-im-internet.de/bundesrecht/abfkl_rv_1992/gesamt.pdf.

17. Bilitewski B, Härdtle G, Marek K: Abfallwirtschaft: Handbuch für Praxis und Lehre. 3rd edition. Berlin; 2000:254–255. ISBN 3-540-64276-5.

18. Länderarbeitsgemeinschaft   Abfall: LAGA-Merkblatt   24: Entsorgung von PCB-haltigen Reststoffen und Abfällen. In Müll Handbuch. Edited by Bilitewski B, Schnurer H, Zeschmar-Lahl B; 1992. Kz. A8375.

19. Umweltbundesamt: PCB-haltige Geräte in Deutschland. http://www.   umweltbundesamt.de/abfallwirtschaft/sonderabfall/pcb.htm, accessed 7.1.2012.

20. EC: Directive 2002/96/EC of the European Parliament and of the Council of 27 January 2003 on waste electrical and electronic equipment (WEEE). OJ L 2003, 37:24.

21. European Environmental Agency: Waste from electric and electronic equipment – quantities, dangerous substances and treatment   methods.   http://eea.eionet.europa.eu/Public/irc/eionet-circle/etc_waste/library?l=/   working_papers/weeepdf/_EN_1.0_&a=d accessed 12.11.2010.

22. Cuhls C: Stoffstrombilanzierung bei der Aufbereitung von Altfahrzeugen und Elektronikschrott in Shredderanlagen. In Abfallvermeidung und – verwertung, Deponietechnik und Altlastensanierung. Edited by Hengerer D, Hofer M, Lorber KE, Ragoßnig A, Nelles M: Balkema; 2000:189–194. ISBN 90 5809 181 3.

23. Deutsche Bundesregierung: Bericht der Bundesregierung zu den abfallwirtschaftlichen Auswirkungen der §§ 9–13 des ElektroG. Berlin: Deutscher Bundestag, BT-Drs; 2011 17/4517.

24. Bezirksregierung Arnsberg. http://www.bezreg-arnsberg.nrw.de/ presse/2010/ 12/204_10/index.php, accessed 7.1.2012.

25. Bezirksregierung   Münster.   http://www.bezreg-muenster. nrw.de/startseite/   presse/pressearchiv/2011/20110216_ Bezirksregierung_Muenster_legt_AGR_   Zwischenlager_ Gelsenkirchen_wegen_PCB_Belastung_still/index.html, accessed 7.1.2012.

26. Bonk L, Böckler M, Göller F, Jasny W, Tigges E: Einsatz, Entsorgung und Recycling PCB-haltiger Bauteile und Komponenten der Elektrotechnik. Gefahrstoffe – Reinhaltung der Luft 2011, 71(1/2):15–19.

27. Landesamt für Natur- und Umweltschutz NRW: Ereignisse und Störfälle in Industrieanlagen. http://www.lanuv.nrw.de/umwelt/ schadensfaelle/anlagen. htm, accessed 19.5.2012.

28. Umweltbundesamt: Emissionen persistenter organischer Schadstoffe. http:// www.umweltbundesamt-daten-zur-umwelt. de/umweltdaten/public/theme. do?nodeIdent=2361, accessed 4.3.2012.

29. Fliedner A, Rüdel H, Jürling H, Müller J, Neugebauer F, Schröder-Kermani C: Levels and trends of industrial chemicals (PCBs, PFCs, PBDEs) in archived herring gull eggs from German coastal regions. Environ Sci Eur 2012, 24:7. doi:10.1186/2190-4715-24-7.

30. Fürst P: Dioxins, PCB, and other organohalogen compounds in human milk. Mol Nutr Food Res 2006, 50:922–933.

31. Bundesinstitut für Risikobewertung: Aufnahme von Umweltkontaminanten über Lebensmittel. Berlin; 2010. ISBN 3-938163-70-4.

32. http://www.umweltprobenbank.de/de/documents/ investigations/results/   analytes?analytes=10011&samplin g_areas=10102&sampling_   years=1982..2010&specimen_ types=10004, accessed 21.11.2010.

33. Moan J, Dahlback A: The relationship between skin cancers, solar radiation, and ozone depletion. Br J Cancer 1992, 65(6):916–921.

34. World Meteorological Organisation: Scientific Assessment of Ozone Depletion: 2010. Geneva: Global Ozone Research and Monitoring Project, Report No. 52; 2011.

35. http://www.engineeringtoolbox.com/refri.

36. UNEP: The Montreal Protocol on Substances that Deplete the Ozone Layer. Nairobi; 2000. ISBN: 92-807-1888-6.

37. UNEP: Decision V/24; http://ozone.unep.org/Publications/MP_ Handbook/MPHandbook-2009.pdf, accessed 13.2.2012.

38. Verordnung zum Schutz des Klimas vor Veränderungen durch den Eintrag bestimmter fluorierter Treibhausgase (ChemikalienKlimaschutzverordnung). 2008, BGBl. I, 1139–1144, 7.7.2008.

39. EC: Regulation No 1005/2009 of the European Parliament and of the Council of 16 September 2009 on substances that deplete the ozone layer. OJ L 2009, 286(1–30):2009.

40. Länderarbeitsgemeinschaft Abfall: Anforderungen zur Entsorgung von Elektronik-Altgeräten. LAGA-Mitteilung 31. www.laga-online.de accessed 26.8.2010.

41. Huisman J: WEEE recast: from 4 kg to 65%: the compliance consequences. Bonn: United Nations University; 2010.

42. Landesamt für Natur-, Umwelt- und Verbraucherschutz NW: Entsorgung FCKW-haltiger Haushaltskühlgeräte in Nordrhein-Westfalen. Recklinghausen: LANUV-Fachbericht 21.

43. Complementaire e-waste stromen in Nederland versie 2.0. http:// www.wecycle. nl/uploads/pdf/onderzoek/2010/2008-07%20 complementaire%20stromen% 20rapport%20v.2%20 W+B.pdf, accessed 28.7.2011.

44. Deutsche Umwelthilfe: Falsche Entsorgung von Kühlschränken setzt FCKW frei. http://www.konsumo.de/news/1919-Umwelt-Falsche-Entsorgung-vonKuehlschraenken-setzt-FCKW-frei, accessed 15.3.2009.

45. Friege H, Schmidt O, Hinken G: Sammlung und Gewinnung von Wertstoffen aus Abfällen – Chancen und Hindernisse. In Abfall-RecyclingAltlasten, Volume 34. Edited by Pinnekamp J. Aachen; 2008. 10/1-10/19.

46. Boeckh M: Kühlmittel erhitzen die Gemüter. Entsorga-Magazin 2011, 7:10–14.

47. E-PRTR: http://prtr.ec.europa.eu/, accessed 17.3.2012.

48. Schwarz R: Emissionen des Kältemittels R134a aus mobilen Klimaanlagen. Frankfurt: Studie im Auftrag des Umweltbundesamts 360 09 006, Öko-Recherche Büro für Umweltforschung und – beratung; 2001.

49. http://www.unep.fr/ozonaction/information/mmcfiles/7508 eOASI2011_TippingtheBalance.pdf, accessed 21.1.2012.

Chapter 3

Scale-Up of the Polyol Process for Nanomaterial Synthesis

Samir Farhat[1], Nassima Ouar[1], Mongia Hosni[1],
Ivaylo Hinkov[2], Silvana Mercone[1], Frédéric
Schoenstein[1], and Noureddine Jouini[1]

[1]Laboratoire des Sciences des Procédés et des Matériaux, CNRS, LSPM-UPR 3407, Université Paris 13, PRES Sorbonne-Paris-Cité, 99 Avenue J.-B. Clément, 93430 Villetaneuse, France

[2]Département de Génie Chimique Université de Technologie Chimique et de Métallurgique, 8 Boulevard St. Kliment Ohridski, 1756 Sofia, Bulgarie

ABSTRACT

Two classes of inorganic materials such as metallic nanowires and metal oxides nanorods were synthesized using the polyol process and scaled-up to produce macroscopic quantities. Scale-up strategy was successfully built by performing the synthesis in a 15 cm diameter, 4.5 liters volume cylindrical tank using a straight paddle impeller and a Rushton turbine. The actual yield of the synthesis is ~45 grams per batch for zinc oxide nanorods and ~20 grams per batch for cobalt nickel nanowires. Under the same rotation speed, the aspect ratio of the produced nanowires and nanorods using the Rushton turbine impeller with radial flow patterns has shown a lower aspect ratio, nanoparticle size and polydispersity. This is attributed to the increase of the local dissipated energy as spatially calculated by computational fluid dynamics (CFD) that is proposed to design, optimize and scale-up the polyol process.

INTRODUCTION

For nanomaterial synthesis, polyol-mediated process is nowadays considered as a very interesting alternative to sol-gel route. While sol-gel route leads to oxides or hydroxides, the polyol-mediated synthesis enables to prepare more classes of inorganic materials (metal, hydroxides, oxides…). It has been established that two main reactions occurred in these media: reduction and hydrolysis, the first leading to metals [1] and the second to hydroxides or oxides [2] . Synthesis of such nanomaterials has gained increasing interest during these two last decades due to technological applications such as information storage [3] or photovoltaic devices [4] . While the majority of these applications rely on precise growth of uniformly sized and shaped nanoparticles, controlled and reproducible synthesis in large scale remains a major challenge which can be addressed through a deeper understanding of nucleation, growth and agglomeration/breakage kinetics connected with heat, mass and momentum transfers in the reaction vessel. As these rate processes are scale dependent, one important task is to find a reliable way to design or model larger scale processes using information obtained from a smaller one. Unfortunately, publications concerning the scale-up of nanomaterials are very sparse in the literature, although its importance has been emphasized by their potential use as advanced materials in the industry. In contrast, classical precipitation from liquid phase in agitated vessels is more used in the chemical industry. Nevertheless, successful scaleup remains difficult due to the absence of a validated methodology that bridges the gap between molecular and macroscopic length scales over a wide range of time scales. Indeed, total similarity on these different scales cannot be fulfilled, as it is difficult or even impossible to maintain constancy in all the dimensionless groups that characterize geometry, thermal and mass transfers, kinetics of the nanoparticle synthesis process. This implies in the particular nanoparticles synthesis conditions that none of the conventional scale-up criteria as equal power input per unit mass, equal tip speed or equal stirring rate is capable of predicting the experimentally observed effects of the mixing conditions on particles size and polydispersity. For simplicity, similarity on different scales could be obtained when the dimensionless groups as Reynolds, Nusselt or Damköhler numbers have the same value on different scales. However, this is difficult to achieve for all the dimensionless groups that characterize the process. In these cases, a trade-off between different dimensionless groups has to be found to maintain geometric, thermal or chemical similarities with scale-up through weighting parameters according to their expected influence on the process. Scale-up then loses its theoretical base and becomes empirical [5] . An alternative approach consists of the mathematical modeling of the underlying physical and chemical processes in the reaction vessel with clearly and quantitatively understood mechanisms, which is not always the case. For these reasons, we built our strategy for scale-up batch precipitation in two steps. In the first step, the synthesis conditions were extrapolated empirically from the laboratory to the pilot scale and validated through ex situ characterizations of the obtained nanomaterials using X-ray diffraction (XRD), high transmission electron microscopy (HTEM), field emission gun scanning electron microscope (FEGSEM), and VSM-SQUID for the magnetic properties. In the second step, and since solving combined computational fluid dynamics (CFD) and population balance (PB) equations still impracticable due to the excessive computational demand and simulation time required [6] [7] , we performed CFD calculations to characterize the turbulent flow patterns and energy dissipation under different axial and radial impellers geometries. The results obtained by these scale-up approaches will be illustrated in this paper on two opposite materials: zinc oxide nanoparticles which is a semi-conducting material used for solar energy conversion based on the photovoltaic effect and metallic cobalt-nickel binary alloy nanowires with potential application in the fields of ultra-high density magnetic storage.

EMPIRICAL SCALE-UP APPROACH

Zinc oxide and $Co_{80}Ni_{20}$ nanoparticles were prepared by extrapolating the stoichiometric conditions from the laboratory to the pilot scale with the intention to maintain the same particle size and morphology. Either axial or radial mixing was tested by using straight paddle and Rushton turbine respectively under the same chemical stoichiometry and the stirring speed. After synthesis, the obtained colloidal suspension containing the residual polyol and the nanoparticles were centrifuged, filtered and cleaned several times and the solid powder dried and characterized by JEOL 2011 transmission electron microscope (TEM) operating at 200 kV. To further confirm phase and morphology, X-ray diffraction was taken in 2q angle between 30° and 80°. The XRD analysis was performed using 2INELTM diffractometers with Cu-Kα1 radiation for ZnO and Co-Kα radiation for Cobaltnickel.

Reactor Description

We designed a new pilot scale reactor that consists of a jacketed glass vessel 4.5 L in volume and 150 mm in internal diameter with double jacket and paddle stirrer. The same reactor was used to produce either zinc oxide or magnetic metals by simply changing the operating conditions. During the synthesis, external reactor surface was insulated from the ambient making the reactor nearly adiabatic. Compared to small lab-scale system, this configuration was designed to improve heat exchange in the reactor, minimizing heat losses since the surface area to volume ratio decreases with volume. In addition, due to the magnetic activity of the cobalt/nickel powder, the steel stirrer was fully encapsulated in Teflon® PTFE. The stirrer paddle was powered by a 50 Watt motor Heidolph RZR 2021 with a digital display and a speed range of 40 to 2000 rpm. As shown in Figure 1, thermal oil circulated through the reactor jacket from a Julabo PrestoLH40 cryo thermostat unit. Internal rector temperature was measured using a Pt100 thermometer with an accuracy of 0.1°C. This parameter was controlled using a self-optimizing controller (ICC—Intelligent Cascade Control) operating in a wide working range from −45°C to 250°C. This allows a rapid heating and cooling and permit temperature stability.

Zinc Oxide Synthesis

Zinc oxide was obtained by forced hydrolysis of zinc acetate salts in diethylene glycol as solvent [8][9] . The chemicals used were zinc acetate dihydrate [(Zn(OAc)$_2$·2H$_2$O)], and diethylene glycol DEG [O(CH$_2$CH$_2$OH)$_2$] from Sigma-Aldrich. Distilled water was added to adjust hydrolysis ratio and ethanol and acetone were used for washing and cleaning nanoparticles. All chemical were of analytical grade without any further purification. To control reaction stoichiometry, we defined the metal concentration in the polyol solution as $z = [Zn^{2+}]$ mol·L^{-1} and the hydrolysis ratio as $h = n_{H_2O} / n_{Zn}^{2+}$ where n_{H_2O} is the number of moles of water including those of zinc acetate dihydrate and n_{Zn}^{2+} accounts for the number of moles of zinc precursor. The reaction was conducted with $h = 5$, $z = 0.5$ at a controlled temperature of 170°C during ~1 hour under continuous stirring at 220 rpm. At these conditions, the solubility of the produced zinc oxide in DEG is very low and high supersaturation is generated which leads to nucleation and crystal growth visible when the solution color's shift from transparent to the white as shown on the photography ofFigure 1(b). After synthesis the white precipitate is washed and drayed.

Cobalt-Nickel Synthesis

For cobalt/nickel nanowires, we used a metal reduction of cobalt and nickel salts introduced into the molar proportions of 80% cobalt and 20% of nickel as discussed in previous works [10] -[15] . Nanowires were obtained by reduction of Cobalt (II) Acetate Tetrahydrate 98% Co(C$_2$H$_3$O$_2$)$_2$·4H$_2$O and Nickel (II) Acetate Tetrahydrate 98%+ Co(C$_2$H$_3$O$_2$)$_2$·4H$_2$O in the presence of Ruthenium (III) Chloride hydrate 99.9% (PGM basis), Ru 38% min (RuCl$_3$·H$_2$O) acting as nucleating agent. The polyol was the 1,2-butanediol (C$_4$H$_{10}$O$_2$) and the base was sodium hydroxide (NaOH). All the chemicals were purchased from Sigma Aldrich. It was found that the basicity of the medium control the morphology of the nanowire [6] . The total concentration of cobalt-nickel in the polyol was 0.08 M, that of sodium hydroxide 0.15 M, and the relative concentration of the nucleating agent [Ru]/[CoNi] = 2.5 × 10^{-2}. For such heterogeneous nucleation process, the number of the formed nuclei and their growth are controlled by [Ru]/[CoNi] ratio. As a consequence, this ratio governs the relative nucleation and growth rates thereby controlling the final nanowire size distribution.

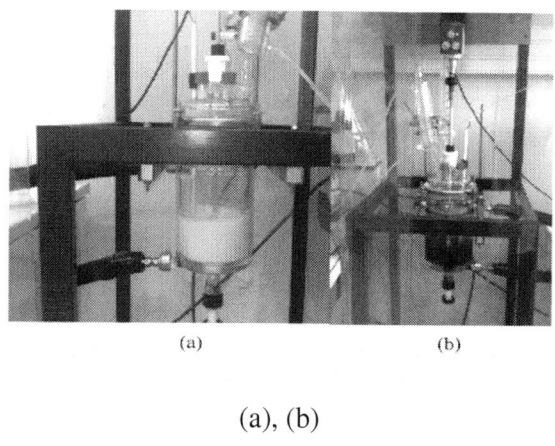

(a), (b)

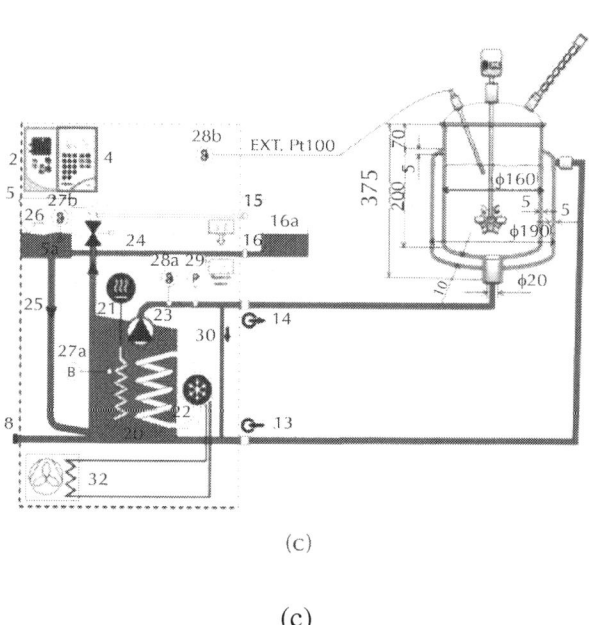

(c)

Figure 1: (a) Schematic of the pilot scale reactor with the overall recirculation heater circuit (grey) and photography of the pilot scale reactor under (b) ZnO and (c) Cobalt Nickel synthesis.

RESULTS

In Figure 2, we report TEM pictures of the zinc oxide nanoparticles obtained under the same stoichiometry with straight paddle (S1) and Rushton turbine (S2) and their corresponding XRD measurements. The total yield of zinc oxide nanoparticles was not affected by the mixing configuration and is ~45 grams of dry material per batch. For small scale reactor the yield is ~1.5 g per batch. TEM pictures confirm the results obtained in small scale where similar ZnO rod-like morphology was obtained. A slight decrease in the nanorod aspect ratio between straight Paddle and Rushton turbine was observed as evidently depicted by the micrographs of Figure 2. The morphological parameters listed in Table 1 were determined from TEM pictures using the ImageJ software program and statistical analysis [16] . XRD patterns of samples S1 and S2 are also shown in Figure 2. They are in agreement with the standard X-ray diffraction peaks, which confirmed that the synthesized materials are wellcrystallized in both mixing situation. In addition, structural analysis confirms our previous results obtained in small scale reactor [16] [17] .

In Figure 3, are shown TEM pictures of the cobalt/nickel nanowires obtained with using straight paddle and Rushton turbine and their corresponding XRD measurements. The total yield of cobalt nickel nanowire was also not affected by the mixing configuration and is ~20 grams of dry material per batch. This yield is lower than zinc oxide synthesis due to different nucleation mechanism and supersaturation conditions. For comparison, the yield obtained at small scale synthesis was less than 1 g per batch. In contrast, a decrease in the nanowire aspect ratio between straight Paddle and Rushton turbine was observed from the statistical analysis of the micrographs of Figure 3. The morphological parameters are summarized in Table1 The X-ray diffraction patterns are showed inFigure 3 where the peaks appear particularly widened suggesting the formation of nano crystalline phases. More specifically, the peak (002) is a signature of a shape anisotropy objects with growth along the crystallographic c axis of the cobalt hcp phase. X-ray diffraction results obtained in pilot scales confirm the structure of the nanowires issued from small scale synthesis and already published [18] [19] .

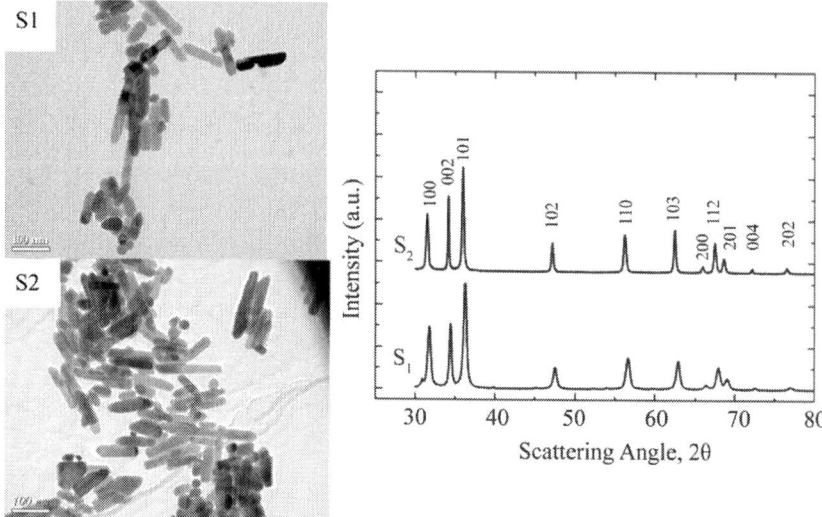

Figure 2: TEM pictures of ZnO nanoparticles obtained with straight paddle (S1) and Rushton turbine (S2) and their respective XRD patterns.

Table 1: Effect of mixing on nanoparticle size obtained in pilot scale. Here L is the nanowire or nanorod length (nm), d the diameter (nm) and L/d the aspect ratio

		Straight paddle	Rushton turbine
Zinc oxide		L = 96 nm	L = 87 nm
		d = 26 nm	d = 28 nm
		L/d = 3.7	L/d = 3.1
Cobalt-nickel		L = 270 nm	L = 170 nm d = 7.6 nm L/d
		d = 6.5 nm L/d = 42	= 22

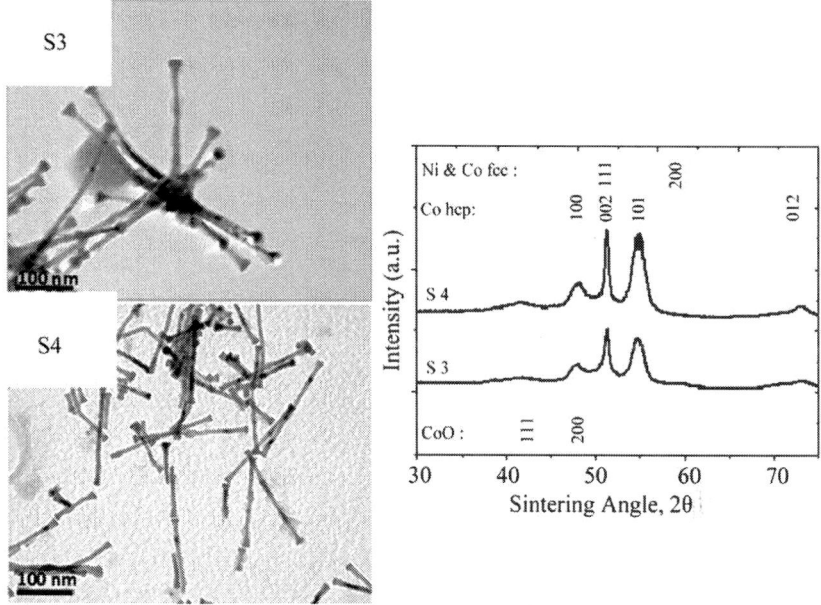

Figure 3: TEM pictures of $Co_{80}Ni_{20}$ nanowires obtained with straight paddle (S3) and Rushton turbine (S4) and their respective XRD patterns.

Since the aspect ratio is crucial for magnetic properties, we compared the magnetic performances of the $Co_{80}Ni_{20}$ samples (S3) and (S4). Hysteresis loops were recorded by vibrating sample magnetometer VSM at room temperature and reported in Figure 4. From these results, we can conclude that the samples show a ferromagnetic behavior characterized by an open cycle, a saturation magnetization (M_s), a remanent magnetization (M_r) and a coercivity (H_c) suggesting that the scale-up is succeeded. Nevertheless, we can notice that the geometry of the paddle has also an effect on the coercive field H_c. When straight paddle was used, the measured coercive field was H_c = 2065 Oe (S3). In contrast, when Rushton turbine (E3) was used, the coercive field was significantly increased to H_c = 3922 Oe. Since the chemical stoichiometry of the reaction was the same, this abnormally elevated coercive field is attributed to the mixing efficiency.

Hence, in order to study the interaction between the bulk liquid phase flow and the tank and paddle geometries, we developed a CFD modeling for the four cases discussed above.

CFD MODELLING

CFD calculations were performed in 3D turbulent fluid flow configuration for different type of agitators and solved using ANSYS FLUENT commercial software [20] . This allows the evaluation of an alternative design and the optimum mixing configuration choose. The fluid around the rotating impeller blades interacts with the stationary baffles and generates a complex, three-dimensional turbulent flow. Other parameters like impeller clearance from the tank bottom, proximity of the vessel walls, baffle length also affect the generated flow. The geometry of the laboratory and pilot scale reactors used in this simulation are depicted schematically in Figure 5 and the geometrical parameters summarized in Table2 For the reactor, a flat-bottom cylindrical tank with diameter DT = 150 mm and baffles was used. In all simulations, the working fluid was diethylene glycol for ZnO synthesis and 1, 2 butanediol for cobalt/nickel nanowires. Temperature dependant viscosity and mass density of these fluids was considered.

To simulate the exact synthesis conditions a volume of 2.65 liters was considered. This corresponds to a total liquid height of HL = DT = 150 mm. in order to avoid vortex formation and improve the mixing efficiency, four equally spaced baffles with width 16 mm and thickness 5.1 mm were placed on the tank wall.

The three dimensional geometry was created using ANSYS Design Modeler and the mesh was generated by ANSYS Meshing application. The impeller region was further refined in order to better capture the flow phenomena occurring there. The impeller rotation was modeled using the Multiple Reference Frame (MRF) approach, where the computational grid comprised two meshes: an inner cylindrical rotating volume enclosing the impeller and an outer stationary volume forming the rest of the reactor including the baffles and separated by an interface. The interface was defined as an imaginary section to provide the interaction between the rotating and stationary frame. Unstructured, non-uniform tetrahedral mesh was generated inside the computational domain. Grids were refined in the wall and impeller regions. After numerous checks for grid sensitivity and mesh constraints, the total number of elements was optimized to provide a computational grid sufficiently fine to resolve the problem within a reasonable CPU time.

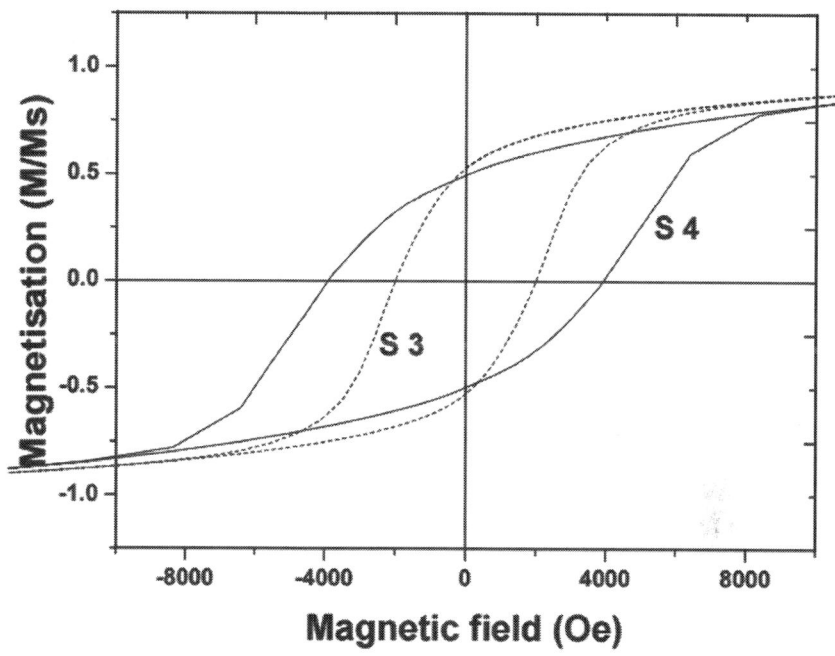

Figure 4: Hysteresis loops of the $Co_{80}Ni_{20}$ powder recorded at room temperature. (S3) Straight paddle. (S4) Rushton turbine.

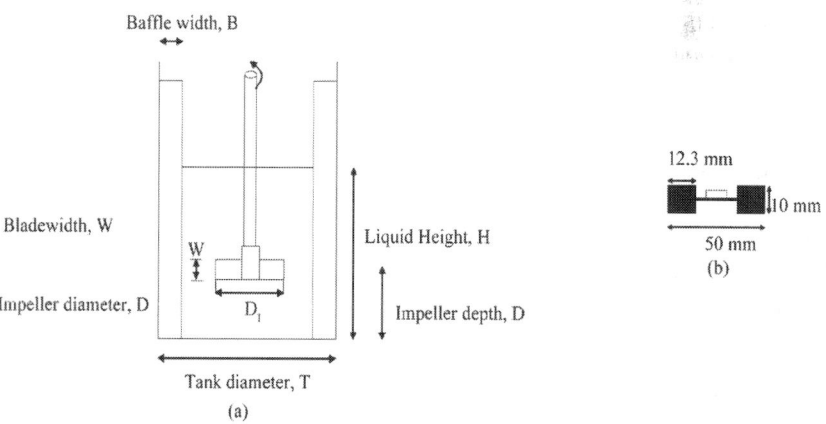

Figure 5: Geometry used in CFD simulation. (a) Straight paddle; (b) Rushton turbine.

Table 2: Geometric characteristics of the impellers

	Straight paddle	Rushton turbine
Diameter	DI = 96 mm	DI = DT/3 = 50 mm
Height of the blade	W = 30 mm	W = 10 mm
Impeller depth	D = 50 mm	D = 50 mm

The hydrodynamic structure of turbulent flow was then simulated using a three-dimensional turbulent model. This model employs a fully conservative finite volume method for the solution of the continuity and momentum equations including turbulence. The governing equations are given in a Reynolds averaged form of the NavierStokes (RANS) equations and summarized in Table3 In Equation (1), U_i is the ith component of the fluid velocity and partial derivatives with respect to xi that are assumed to one of the three coordinate directions. When turbulence is included, the velocity is assumed to be the sum of equilibrium and fluctuating components, U_i and u_i' respectively. In Equation (2), the temporal and convection terms are on the left. Whereas the terms on the right hand side are the pressure gradient; the divergence of the stress tensor, which is responsible for the diffusion of momentum; the Reynolds stresses, involving the terms (the over bar indicates that these terms are timeaveraged values) and the gravitational force, respectively.

Table 3: Model equations

$$\frac{\partial \rho}{\partial t} + \frac{\partial (\rho U_i)}{\partial x_i} = 0 \tag{1}$$

$$\frac{\partial (\rho U_i)}{\partial t} + \frac{\partial}{\partial x_j}(\rho U_i U_j) = -\frac{\partial P}{\partial x_i} + \frac{\partial}{\partial x_j}\left[\mu\left(\frac{\partial U_i}{\partial x_j} + \frac{\partial U_j}{\partial x_i} - \frac{2}{3}\frac{\partial U_k}{\partial x_k}\delta_{ij}\right)\right] + \frac{\partial}{\partial x_j}\left(-\overline{\rho u_i' u_j'}\right) + \rho g_i \tag{2}$$

$$\frac{\partial (\rho k)}{\partial t} + \frac{\partial (\rho U_i k)}{\partial x_i} = \frac{\partial}{\partial x_i}\left[\left(\mu + \frac{\mu_t}{\sigma_k}\right)\frac{\partial k}{\partial x_i}\right] + G_k - \rho \varepsilon \tag{3}$$

$$\frac{\partial(\rho\varepsilon)}{\partial t} + \frac{\partial(\rho U_i\varepsilon)}{\partial x_i} = \frac{\partial}{\partial x_i}\left[\left(\mu + \frac{\mu_t}{\sigma_\varepsilon}\right)\frac{\partial\varepsilon}{\partial x_i}\right] + C_{1\varepsilon}\frac{\varepsilon}{k}G_k + C_{2\upsilon}\,\rho\frac{\varepsilon^2}{k}$$

(4)

$$G_k = \mu_t\left(\frac{\partial U_i}{\partial x_j} + \frac{\partial U_j}{\partial x_i}\right)\left(\frac{\partial U_j}{\partial x_i}\right)$$

(5)

$$\mu_t = \rho C_\mu\frac{k^2}{\varepsilon}$$

(6)

The governing equations in the flow region surrounding the impeller were modified for the rotating frame. In steady RANS, the time-averaged velocity is resolved and the velocity fluctuations are modeled through turbulence models. The selection of the turbulence approach is critical to fully resolving the fundamentals of the mixing behavior. For solving the Reynolds stresses, the standard k–ε model with enhanced wall treatment was used. This model assumes that the normal stresses are roughly equal and are adequately represented by the turbulent kinetic energy. Two transport equations are used to model the production, distribution, and dissipation of turbulent kinetic energy: the k-Equation (3), and the ε-Equation (4). In Equations (3) and (4), G_k represents the generation of turbulence kinetic energy due to the mean velocity gradients given by Equation (5) where $C_1\varepsilon$, $C_2\varepsilon$, σ_k and $\sigma\varepsilon$ are empirical constants. The turbulent or eddy viscosity, μt is computed by combining k and ε by Equation (6).

The model constants used in our simulations are: $C_1\varepsilon$ = 1.44; $C_2\varepsilon$ = 1.92; C_μ = 0.09; σ_k = 1.0 and $\sigma\varepsilon$ = 1.3. For initial and boundary conditions, cylindrical wall, bottom wall, and baffles were modeled as stationary, impermeable walls. On these walls a no-slip condition was applied for the liquid. The impeller shaft and hub or discs, in the case of Rushton turbine, were specified as moving walls with angular velocity corresponding to the impeller rotational speed. A free surface boundary condition was defined at the liquid surface. The operating pressure of 1 atm was set at the liquid surface on the top of the tank. As a first approximation, only the pure solvent diethylene glycol or 1,2 Butanediol was considered with temperature dependent mass density μ and dynamic viscosity μ. The impeller shaft and the gravity acceleration were defined along the vertical axis. The impeller rotating speed is set to 220 rpm in all cases. The flow field was calculated using a steady state assumption. The velocity-pressure coupling was achieved using SIMPLE algorithm to iteratively solve the discretized equations. First order upwind discretization scheme was adopted for pressure interpolation and the convective term of momentum and turbulent kinetic energy. The solution residuals for all the conservation Equations (2)-(4) were set to a value of 5×10^{-5}. Figure 6 gives the velocity magnitude obtained by the simulation during ZnO synthesis with straight paddle (S1) and Rushton turbine (S2). Similar behaviors were obtained with experiments (S3) and (S4) and are not reported here. For the straight paddle the flow is axial like and the maximum velocity is obtained very close to the impeller at a value of ~1.7 m/s. In contrast, with Rushton turbine, the flow patterns are radial like with a maximum velocity of ~0.8 m/s.

Figure 7 shows the typical contours of the Reynolds number and dissipation of kinetic energy visualized in the transversal plane containing the paddle. The flow field produced by the Rushton turbine is characterized by higher local power dissipation in the impeller zone with the discharge stream of the impeller being mainly radial. By contrast, the straight-type impeller produces a fairly uniform energy distribution throughout the vessel. From these results, two important parameters appear to affect particle size distribution: the flow rates of reagents or Reynolds number (R_e) and the energy dissipation (e). Increasing both of them permits to reduce particle size and decreases the distribution width in agreement with several published works [21] -[23] .

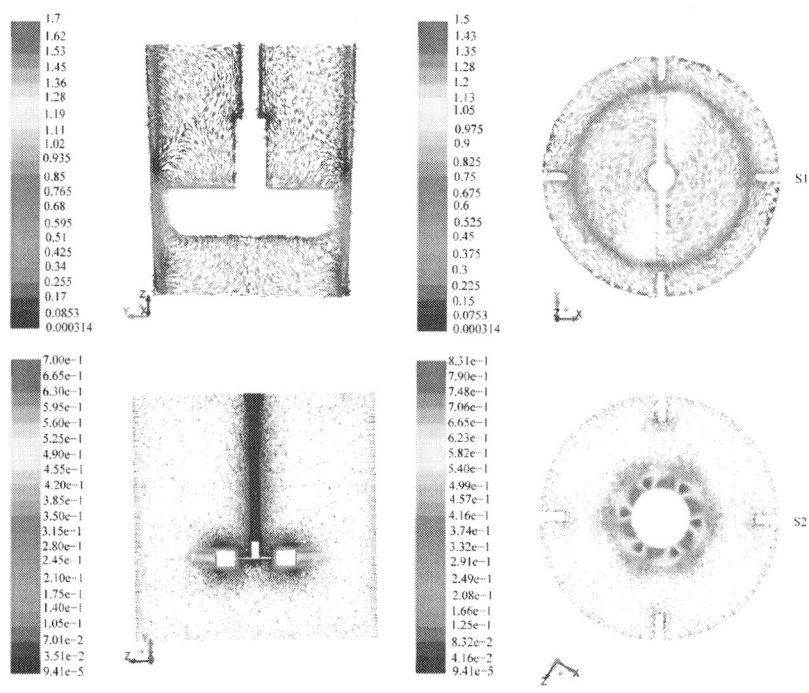

Figure 6: Velocity magnitude in m·s⁻¹ depicting the flow features in a vertical and horizontal planes in pilot scale with axial (S1) and radial (S2) mixing. Conditions are diethylene glycol as solvent, N = 220 RPM and T = 170°C.

CONCLUSIONS

As a conclusion, this study permitted to demonstrate that it is possible to reproduce at larger scale reactors the auspicious conditions for both forced hydrolysis and reduction of metallic salts in polyol medium. We performed the scale-up of these two distinct classes of inorganic nanomaterials in forms of metals and oxides with sample aliquots actually in the tens gram scale. This is useful to elaborate fairly homogenous samples issued from the same batch as already demonstrated in our recent works for photovoltaic applications of zinc oxide [16] and spark plasma sintering (SPS) of cobalt nickel magnetic nanowires [18] [19] .

From fundamental point of view, successful scale-up is possible if shear rate distribution in the reactor is precisely controlled during the synthesis. Nevertheless, the high viscosity and low diffusion coefficients of metallic ions in polyols strongly affect mass transport and reaction kinetics. In order to avoid this negative effect, mixing systems must be optimized to facilitate the reactions by increasing the local dissipated energy. Consequently, it is possible to control local synthesis conditions by choosing the appropriate mixing system that produce nanoparticles with a narrow size distribution. As performed in this paper, CFD calculations could help to depict the turbulent flow patterns and energy dissipation under different impellers geometries giving a better picture of the mixing efficiency in the reaction vessel. Indeed, mixing, which results in bulk movement of the fluid, plays a significant role in maintaining the homogeneity of a given reactive system as well as heat and mass transfers. In the context of large-size reactors, good mixing is especially important since it is essential to maintain the same rate of the mass transfer in to the overall reaction medium thereby insuring uniformly sized and shaped nanoparticles. The richness of the polyol process associated with chemical engineering methodology could open interesting perspectives for the mass synthesis of several classes of nanomaterials [24] [25] as well as the design of their reactor synthesis from academic-level prototyping to industrial-level manufacturing.

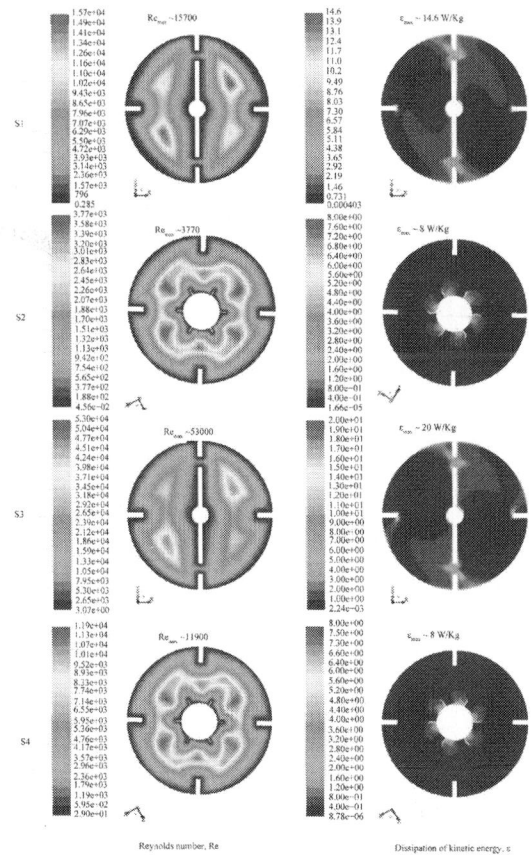

Figure 7: Simulated contours of Reynolds number (Re) and turbulent dissipation of kinetic energy (ε) for ZnO (S1 and S2) and $Co_{80}Ni_{20}$ (S3 and S4) synthesis. Conditions are DEG at 170°C for ZnO and 1,2 Butanediol at 165°C for $Co_{80}Ni_{20}$. Rotation speed N = 220 RPM.

ACKNOWLEDGEMENTS

IFR Paris Nord Plaine de France (PPF) as well as ANR (Agence Nationale de la Recherche) and CGI (Commissariat à l'Investissement d'Avenir) are gratefully acknowledged for their financial support of this work through Labex SEAM (Science and Engineering for Advanced Materials and Devices) ANR 11 LABX 086, ANR 11 IDEX 05 02.

REFERENCES

1. Figlarz, M., Fiévet, F. and Lagier, J.P. (1985) French Patent No. 8221483.

2. Poul, L., Ammar, S., Jouini, N., Fievet, F. and Villain, F. (2003) Synthesis of Inorganic Compounds (Metal, Oxide and Hydroxide) in Polyol Medium: A Versatile Route Related to the Sol-Gel Process. Journal of Sol-Gel Science and Technology, 26, 261-265.http://dx.doi.org/10.1023/A:1020763402390

3. Whitney, T.M., Searson, P.C., Jiang, J.S. and Chien, C.L. (1993) Fabrication and Magnetic Properties of Arrays of Metallic Nanowires. Science, 261, 1316-1319.http://dx.doi.org/10.1126/science.261.5126.1316

4. Keis, K., Bauer, C., Boschloo, G., Hagfeldt, A., Westermark, K., Rensmo, H. and Siegbahn, H. (2002) Nanostructured ZnO Electrodes for Dye-Sensitized Solar Cell Applications. Journal of Photochemistry and Photobiology A: Chemistry, 148, 57-64. http://dx.doi.org/10.1016/S1010-6030(02)00039-4

5. Zauner, R. (1994) Scale-Up of Precipitation Processes. Ph.D. Thesis, University of London, London.

6. Gradl, J., Schwarzer, H.-C., Schwertfirm, F., Manhart, M. and Peukert, W. (2006) Precipitation of Nanoparticles in a T-Mixer: Coupling the Particle Population Dynamics with Hydrodynamics through Direct Numerical Simulation. Chemical Engineering and Processing: Process Intensification, 45, 908-916.http://dx.doi.org/10.1016/j.cep.2005.11.012

7. Schwarzer, H.-C., Schwertfirm, F., Manhart, M., Schmid, H.-J. and Peukert, W. (2006) Predictive Simulation of Nanoparticle Precipitation Based on the Population Balance Equation. Chemical Engineering Science, 61, 167-181.http://dx.doi.org/10.1016/j.ces.2004.11.064

8. Jézéquel, D., Guenot, J., Jouini, N. and Fiévet, F. (1995) Submicrometer Zinc Oxide Particles: Elaboration in Polyol Medium and Morphological Characteristics. Journal of Materials Research, 10, 77-83. http://dx.doi.org/10.1557/JMR.1995.0077

9. Dakhlaoui, A., Jendoubi, M., Smiri, L.S., Kanaev, A. and Jouini, N. (2009) Synthesis, Characterization and Optical Properties of

ZnO Nanoparticles with Controlled Size and Morphology. Journal of Crystal Growth, 311, 3989-3996.http://dx.doi.org/10.1016/j.jcrysgro.2009.06.028

10. Ung, D., Soumare, Y., Chakroune, N., Viau, G., Vaulay, M.J., Richard, V. and Fiévet, F. (2007) Growth of Magnetic Nanowires and Nanodumbbells in Liquid Polyol. Chemistry of Materials, 19, 2084-2094. http://dx.doi.org/10.1021/cm0627387

11. Soumare, Y., Piquemal, J.Y., Maurer, T., Ott, F., Chaboussant, G., Falqui, A. and Viau, G. (2008) Oriented Magnetic Nanowires with High Coercivity. Journal of Materials Chemistry, 18, 5696-5702. http://dx.doi.org/10.1039/b810943e

12. Soumare, Y., Garcia, C., Maurer, T., Chaboussant, G., Ott, F., Fiévet, F., Piquemal, J.-Y. and Viau, G. (2009) Kinetically Controlled Synthesis of Hexagonally Close-Packed Cobalt Nanorods with High Magnetic Coercivity. Advanced Functional Materials, 19, 1971-1977.http://dx.doi.org/10.1002/adfm.200800822

13. Viau, G., Garcia, C., Maurer, T., Chaboussant, G., Ott, F., Soumare, Y. and Piquemal, J.Y. (2009) Highly Crystalline Cobalt Nanowires with High Coercivity Prepared by Soft Chemistry. Physica Status Solidi A, 206, 663-666.

14. Liu, Q., Guo, X., Wang, T., Li, Y. and Shen, W. (2010) Synthesis of CoNi Nanowires by Heterogeneous Nucleation in Polyol. Materials Letters, 64, 1271-1274.http://dx.doi.org/10.1016/j.matlet.2010.03.006

15. Ait Atmane, K., Zighem, F., Soumare, Y., Ibrahim, M., Boubekri, R., Maurer, T., Margueritat, J., Piquemal, J.-Y., Ott, F., Chaboussant, G., Schoenstein, F., Jouini, N. and Viau, G. (2013) High Temperature Structural and Magnetic Properties of Cobalt Nanorods. Journal of Solid State Chemistry, 197, 297-303.http://dx.doi.org/10.1016/j.jssc.2012.08.009

16. Hosni, M., Kusumawati, Y., Farhat, S., Jouini, N. and Pauporté, T. (2014) Effects of Oxide Nanoparticle Size and Shape on Electronic Structure, Charge Transport and Recombination in Dye-Sensitized Solar Cell Photoelectrodes. The Journal of Physical Chemistry C, 118, 16791-16798.http://dx.doi.org/10.1021/jp412772b

17. Hosni, M., Farhat, S., Schoenstein, F., Karmous, F., Jouini, N., Viana, V. and Mgaidi, A. (2014) Ultrasound Assisted Synthesis of Nanocrystalline Zinc Oxide: Experiments and Modelling. Journal

of Alloys and Compounds, in Press.http://dx.doi.org/10.1016/j.jallcom.2013.12.056.

18.   Ouar, N., Bousnina, M.A., Schoenstein, F., Mercone, S., Brinza, O., Farhat, S. and Jouini, N. (2014) Spark Plasma Sintering of Co80Ni20 Nanopowders Synthesized by Polyol Process and Their Magnetic and Mechanical Properties. Journal of Alloys and Compounds, in press. http://dx.doi.org/10.1016/j.jallcom.2014.01.058.

19.   Ouar, N., Schoenstein, F., Mercone, S., Farhat, S., Villeroy, B., Leridon, B. and Jouini, N. (2013) Spark-Plasma-Sintering Magnetic Field Assisted Compaction of Co80Ni20 Nanowires for Anisotropic Ferromagnetic Bulk Materials. Journal of Applied Physics, 114, Article ID: 163907. http://dx.doi.org/10.1063/1.4827199

20.   Fluent Commercial Software, ANSYS® Version 13.0.

21.   Bockhorn, H., Mewes, D., Peukert, W. and Warnecke, H.-J. (2010) Micro and Macro Mixing Analysis, Simulation and Numerical Calculation. Springer-Verlag Berlin Heidelberg. http://dx.doi.org/10.1007/978-3-642-04549-3

22.   Winkelmann, M., Schuler, T., Uzunogullari, P., Winkler, C.A., Gerlinger, W., Sachweh, B. and Schuchmann, H.P. (2012) Influence of Mixing on the Precipitation of Zinc Oxide Nanoparticles with the Miniemulsion Technique. Chemical Engineering Science, 81, 209-219. http://dx.doi.org/10.1016/j.ces.2012.06.036

23.   Bensaid, S., Deorsola, F.A., Marchisio, D.L., Russo, N. and Fino, D. (2014) Flow Field Simulation and Mixing Efficiency Assessment of the Multi-Inlet Vortex Mixer for Molybdenum Sulfide Nanoparticle Precipitation. Chemical Engineering Journal, 238, 66-77.http://dx.doi.org/10.1016/j.cej.2013.09.065

24.   He, B., Chen, Y., Liu, H. and Liu, Y. (2005) Synthesis of Solvent-Stabilized Colloidal Nanoparticles of Platinum, Rhodium and Ruthenium by Microwave-Polyol Process. Journal of Nanoscience and Nanotechnology, 5, 266-270.http://dx.doi.org/10.1166/jnn.2005.028

25.   Jo, Y.H., Park, J.C., Bang, J.U., Song, H. and Lee, H.M. (2011) New Synthesis Approach for Low Temperature Bimetallic Nanoparticles: Size and Composition Controlled Sn-Cu Nanoparticles. Journal of Nanoscience and Nanotechnology, 11, 1037-1041.http://dx.doi.org/10.1166/jnn.2011.3052

Chapter 4

Recent Advances of Modern Protocol for C-C Bonds—The Suzuki Cross-Coupling

Jadwiga Sołoducho, Kamila Olech, Agnieszka Świst, Dorota Zając, Joanna Cabaj

Faculty of Chemistry, Wrocław University of Technology, Wrocław, Poland

ABSTRACT

Over the past 20 years, small molecule solid phase synthesis has become a powerful tool in the discovery of novel molecular materials. In the development of organic chemistry, the carbon-carbon bond formation has always been one of the most useful and fundamental reaction. The current review summarizes recent developments in metal-catalyzed coupling reactions. The following method is discussed in detail—the cross-coupling of aryl halides with aryl boronic acids (the Suzuki coupling), and the others C-C bond formation reactions as the palladium-catalyzed reaction between an aryl and (or) alkyl halide and a vinyl functionality (the Heck reaction); and the palladium-catalyzed cross-coupling reaction of organostannyl reagents with a variety of organic electrophiles (the Stille reaction)—are mentioned.

INTRODUCTION

Recently, small-molecule solid-phase synthesis has become a powerful tool in the discovery of new drugs or molecular materials. The methods are investigated by the time in details: the cross-coupling of aryl halides with aryl boronic acids (the Suzuki coupling); the palladiumcatalyzed reaction between an aryl and (or) alkyl halide and a vinyl functionality (the Heck reaction); and the palladium-catalyzed cross-coupling reaction of organostannyl reagents with a variety of organic electrophiles (the Stille reaction) [1].

Organic transformations in aqueous media have received also much attention because water is mainly harmless to the environment. The cross-coupling reaction of alkenyl and aryl halides with organoborane derivatives in the presence of a palladium catalyst and a base (Suzuki reaction) has often been carried out in an organicaqueous mixed solvent. Although Davidson and Triggs [2] discovered in 1968 that arylboronic acids reacted with palladium(II) acetate to give corresponding biaryls, and Garves [3] in 1970 that arylsulfinic acids could be coupled to biaryls using Pd(II) in aqueous solvents, it was not until 1979 when biaryls could efficiently be prepared by a palladium-catalyzed reaction. Miyaura and Suzuki [4] reported that cross-coupling reactions between alkenylboranes and organic halides were efficiently catalyzed by a catalytic amount of tetrakis (triphenylphoshine) palladium $(Pd(PPh_3)_4)$ in the presence of a suitable base.

In this context, using the conventional bridged derivatives of phenoxazine [5], phenothiazine, fluorene, carbazole as a functional segments, the palladium-catalyzed condensation halide arylenes by Suzuki-, and other coupling reactions were used as well-established procedure in modern organic synthesis [6]. The coupling products find good applications as intermediates in the preparation of materials (Figure 1). However, the Suzuki-type reactions catalyzed by heterogeneous palladium involving bromoarenes are relatively unexplored. In this context, polymer-supported catalysts have become valuable tools offering many advantages such as simplification of product synthesis.

Nowadays, palladium-catalyzed Suzuki-Miyaura coupling reaction [7,8] of aryl halides is one of the most effective methods for the construction of biaryls or substituted aromatic moieties in organic synthesis. Although homogeneous palladium catalysts have been extensively investigated [9-14], their industrial applications remained limited due to the difficulty in the separation process from the products for recycling [9-15]. Heterogenization of the existing homogeneous palladium catalysts offers an attractive solution to this problem [16-19]. From a practical point of view, the use of aryl chlorides is highly desirable compared with the more expensive aryl iodides, aryl bromides and aryl triflates.

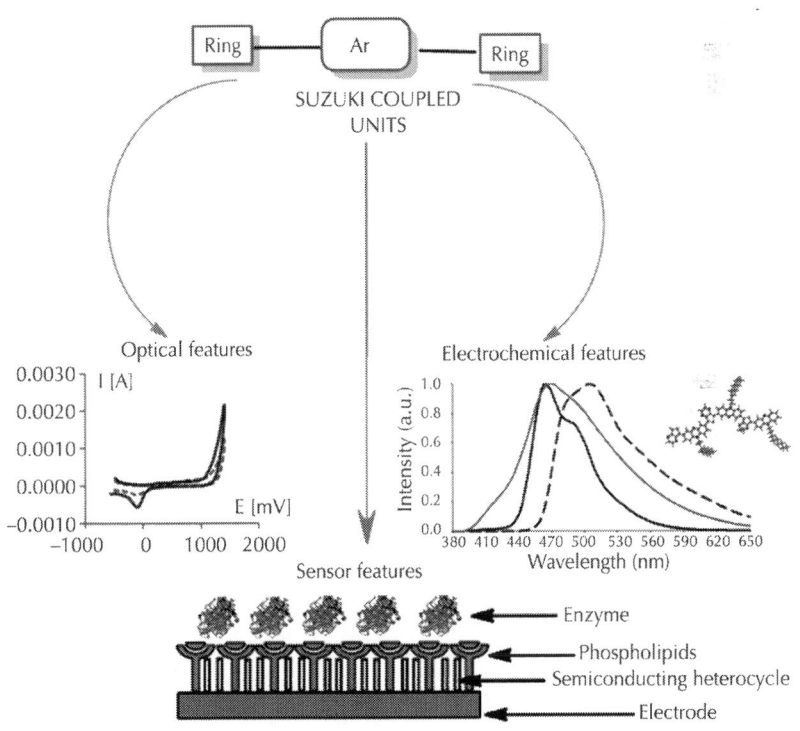

Figure 1: Applications of molecular materials synthesized by Suzuki coupling.

This short review summarizes the advantages of the Suzuki protocol over other similar cross-coupling reactions, several structurally different ligands that promote the reaction in water [20-23], recent results from ligandless Suzuki procedure, and target molecules mainly produced by the method.

TRANSITION METALS IN SYNTHETIC ORGANIC CHEMISTRY

During the second half of the 20th century, transition metals have come to play an important role in organic chemistry and this has led to the development of a large number of transition metal-catalyzed reactions for creating organic molecules. Transition metals have a unique ability to activate various organic compounds and through this activation they can catalyze the formation of new bonds. One metal that was used early on for catalytic organic transformations was palladium. One event that stimulated research into the use of palladium in organic chemistry was the discovery that ethylene is oxidized to acetaldehyde by air in a palladium-catalyzed reaction and this became the industrially important Wacker process. Subsequent research on palladium-catalyzed carbonylation led to new reactions for the formation of carbon-carbon bonds.

In general, transition metals, and in particular palladium, have been of importance for the development of reactions for the formation of carbon-carbon bonds. In 2005 the Nobel Prize in chemistry was awarded to metal-catalyzed reactions for the formation of carbon-carbon double bonds. The 2010 year the Nobel Prize in chemistry is awarded to the formation of carbon-carbon single bonds through palladium-catalyzed cross-coupling reactions.

SUZUKI COUPLING

The Scheme 1 shows the first published Suzuki coupling, which is the palladium-catalyzed cross-coupling between organoboronic acid and halides. Recent catalyst and methods developments have broadened the possible applications enormously, so that the scope of the reaction partners is not restricted to aryls, but includes alkyls, alkenyls and alkynyls. Potassium trifluoroborates and organoboranes or boronate esters may be used in place of boronic acids. Some pseudohalides (for example triflates) may also be used as coupling partners.

MECHANISM OF THE SUZUKI COUPLING

In the Suzuki (Scheme 2) as well as Stille mechanism is need of activation, and in the Suzuki coupling the boronic acid must be activated, for example with base. The activation of the boron atom features the polarization of the organic ligand, and facilitates transmetalation. If starting materials are substituted with base labile groups (i.e., esters), powdered KF effects this activation while leaving base labile groups unaffected. It is worth to mention, that Fis also advantageous in Stille reaction.

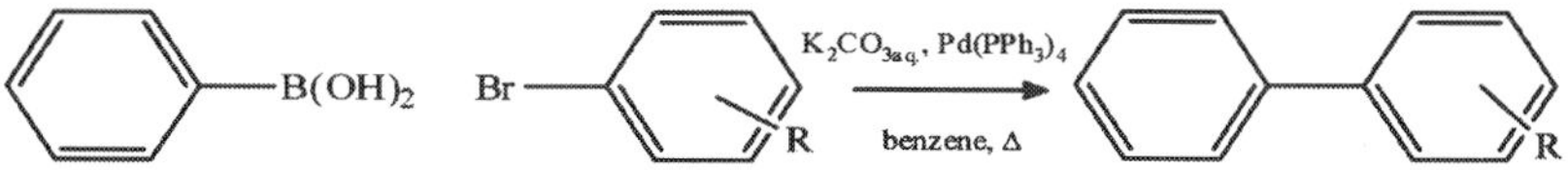

Scheme 1: Typical Suzuki procedure.

Due to the stability, ease of preparation and low toxicity of the boronic acid compounds, there is currently widespread interest in applications of the Suzuki coupling, with new developments and refinements being reported constantly.

Oxidative Addition

Attaching of the palladium catalyst to the alkyl halide gives rise to the organopalladium complex (Scheme 3). The complex is at the beginning in the cis conformation but isomerizes to the trans conformation. Stereochemistry with vinyl halides are retained but inversion of stereochemistry occurs with allylic or benzylic halides.

Transmetalation

The role of base in the Suzuki coupling is to activate the boron-containing reagent, and also facilitate the forma- tion of $R^1Pd\text{-}OR$ from $R^1Pd\text{-}X$ (Scheme 4). Reaction does not occur in the absence of base. Exact mechanism is unclear.

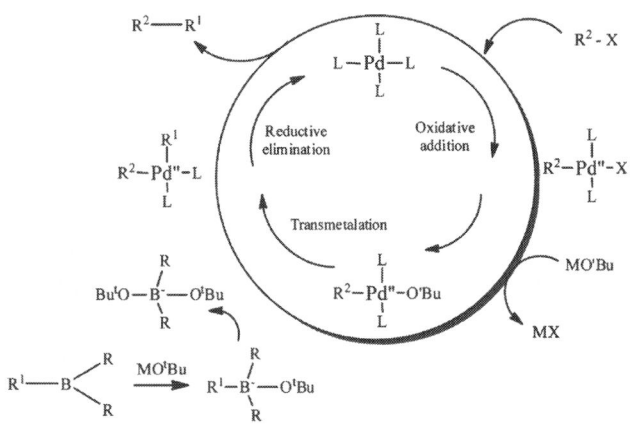

Scheme 2: The mechanism of Suzuki coupling reaction.

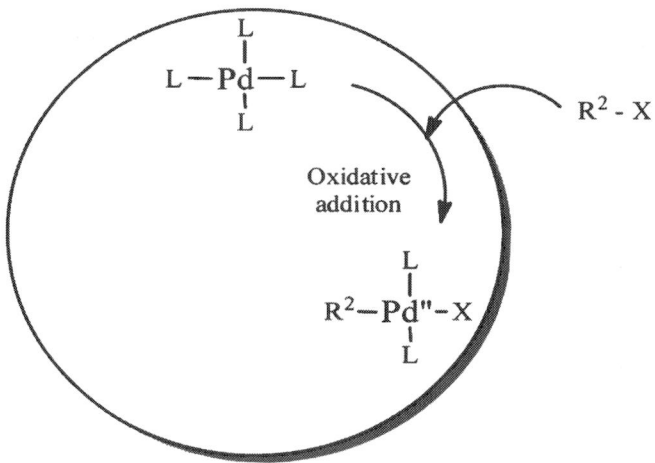

Scheme 3: Oxidative addition of the catalytic cycle.

Reductive Elimination

This final step of catalytic cycle gives the desired product and it also regenerates the palladium catalyst so that it can participate again in the catalytic cycle (Scheme 5). Require the complex to revert back to the cis conformation before reductive elimination can occur.

SUZUKI CROSS-COUPLING AS SYNTHETIC PROTOCOL FOR BRANCHED HETEROCYCLIC UNITS

Suzuki reactions are very useful tool for the formation of Csp_2-Csp_2 bonds. This procedure is, however, convenient in the synthesis of conducting polymers. Suzuki reaction is also a powerful method for the formation of Csp_2-Csp_3 bond. This coupled protocols was also used in our previous work for obtaining branched or hyper-branched sterically crowded heterocyclic structures [24,25] (Scheme 6).

Phenothiazine, as well as phenoxazine, is a wellknown very strong electron donor and has high HOMO energy level because of sulfur atom. Therefore, it was expected phenothiazine ring to be an excellent building block for lowering the ionization potential of conjugated polymers. However, efficient electrogenerated chemiluminescence was observed in co-oligomers of methyl- phenothiazine and interesting redox properties were found in other oligophenothiazines [26]. Phenothiazines as well as phenoxazines are nonplanar. The possible consequence of the nonplanarity of these rings for the photophysics, light-emitting properties, charge transport of ϖ-conjugated polymers are motivated. However, the device based on poly(alkylphenothiazine) are not enough to make efficient device since it has unbalanced charge transporting property. In order to improve electron affinity and transporting properties of material we decided to introduced the benzothiadiazole to the oligomer backbone.

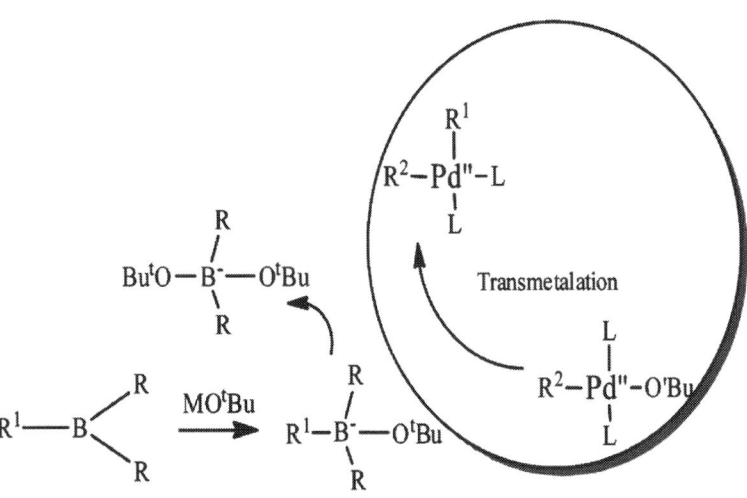

Scheme 4: Transmetalation mechanism.

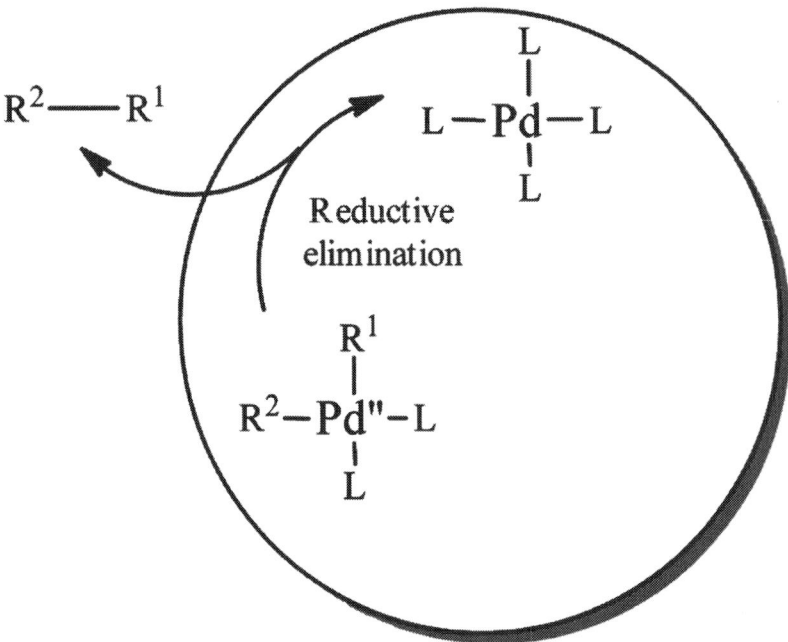

Scheme 5: Reductive elimination.

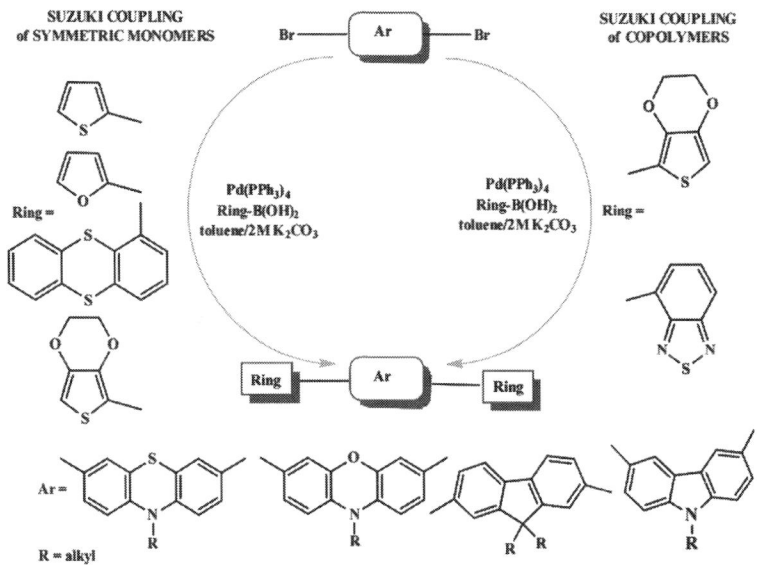

Scheme 6: Pathway of synthesis of symmetric derivatives of arylenes.

Suzuki coupling of dibromoarylenes and boronic acid derivatives was subsequently performed using Pd(PPh$_3$)$_4$ as the catalyst. The synthesis of phenothiazine/phenoxazine oligomers was provided as shown in Scheme 7. This type of reactions was carried out in standard conditions [24].

Conversion of N-alkyl-3,7-dibromophenoxazine, -phenothiazine to the boronic ester was achieved under reaction conditions in presence 2-isopropoxy-4,4,5,5-tetramethyl- 1,3,2-dioxaborolane [24]. Alternated benzothiadiazole oligomers were obtained in 50% yield. The molar ratio of phenoxazine moiety in the oligomers was controlled by adjusting the molar ratio between dibromobenzothiadiazole and monomers ab while a 1/1 molar ratio between the dibromides and the bisborylated compounds was maintained [24].

It was determined that the oligomers were all of rather low molecular weight. Both gave number-average molecular weights of about 3500 g/mol. This value corresponds to a degree of polymerization of seven to eight, meaning that the polymers synthesized here have an average of about 30 aromatic rings per chain [24]. The molecular weights obtained were adequate for processing and film formation, and thus, further optimization of the polymerization was not attempted.

2,1,3-Benzothiadiazole-based oligomers and polymers have been widely studied in recent years as active materials in various optoelectronic devices because of the heterocyclic group and the observed low-band-gap in polymers containing it [27]. Copolymerization of benzothiadiazole with phenoxazine, phenothiazine, fluorene [27] and carbazole [28] or other suitable arylenes can be used as a means to tune the HOMO-LUMO levels in the resulting polymers.

The electron-rich, heterocyclic thianthrene seemed to be a good example for charge transporting materials, since it shows a reversible oxidation behavior at low potential in cyclic voltammetry. The synthesis of thianthrene derivatives (Scheme 8) in moderate yield is shown as follows [25]. The cross-coupling reaction of thianthren-1-yl boronic acid with dibromoalkyl derivatives of fluorene-, carbazole-, diphenylamine-, phenothiazine in presence of catalytic amounts of Pd $(PPh_3)_4$ in two-phases system at 90°C gave a-d, respectively. The obtained semiconducting units as viable luminance and high hole-transporting one, exhibit good solubility in common organic solvents, thermal stability and luminescence in blue region and can be cast into uniform films. They possess good fluorescence quantum [25].

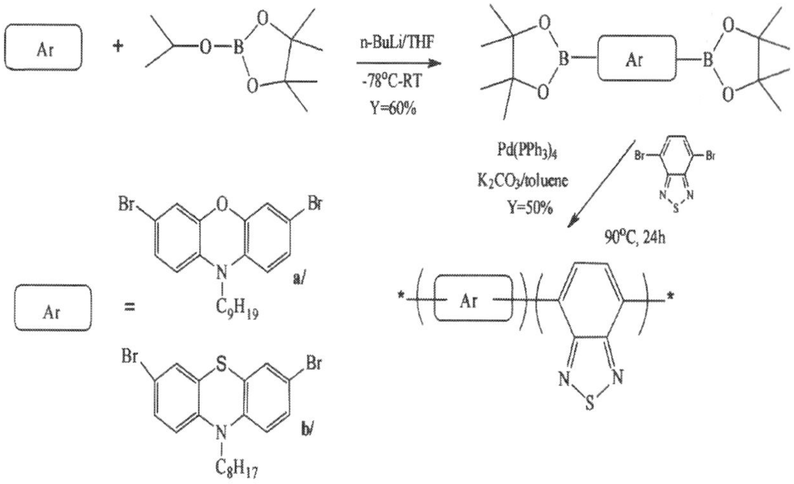

Scheme 7: Synthetic route to the co-oligomers based on benzothiadiazole [by 24].

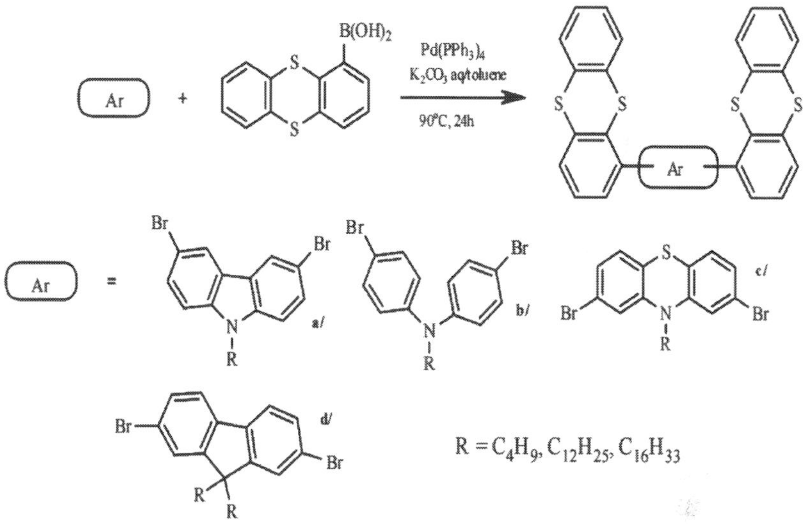

Scheme 8: The synthetic procedure for Suzuki coupled thianthrene derivatives [by 25].

SUZUKI CROSS-COUPLING VS. OTHER PALLADIUM-CATALYZED REACTIONS

In the Suzuki reaction [8,29], the Heck reaction [30], the Kumada reaction [31], the Stille reaction [32], the Negishi reaction [33], and the Sonogashira reaction [34], palladium is known to be particularly effective in activating sp_2-carbon-halogen bonds even in aqueous media [35]. All reaction types have drawbacks that limit the use in synthesis (Table 1).

Suzuki cross coupling has, on the other hand, less limitations than the other reactions mentioned. In the Heck reaction, for example, where an aryl or vinyl halide and an alkene are converted to a more highly substituted alkene under alladium catalysis, the intermolecular reaction often proceeds well when the alkene is electrophilic. With nucleophilic substituents, the reaction gives less satisfactory effects. The Kumada coupling is very sensitive to air and the presence of radical inhibitors, and this has limited the use of the reaction in aqueous media. In the Stille reaction, stannanes are used as substrates, and many of these are environmentally hazardous. There is no toxicity issue involved in organoborane reagents. For a large-scale setting, a Suzuki coupling is an attractive choice. It is assumed that the electron-donating character of the alkyl groups on the phosphine, as well as its steric bulk, results in a more facile oxidative addition of palladium into the aryl halide. Prior work utilizing aryl chlorides in Suzuki couplings required strongly electronwithdrawing substituents on the aryl chloride.

The organoborane reagents are usually more easy to prepare in the laboratory, and the workup procedures are more simple than using other cross-coupling methods [27-40].

LIGANDS SUZUKI CROSS-COUPLING

The vast majority of ligands applied in modern synthesis possess low water solubility. Many strategies have been used to enhance the solubility of metal-ligand complexes in water of which three are mostly utilized [41]. The catalytic reaction is performed in the presence of micelle-forming surfactants, solubilizing functionalities are added to the ligand, i.e., ionic groups to poorly soluble ligands and easily accessible, water-soluble sources from nature are utilized as ligands.

Table 1. Comparison of palladium catalyzed reactions

Type of coupling	Reaction
The Negishi Coupling	R^1-ZnR^2 + R^2-X $\xrightarrow{\text{cat. }[Pd^0L_n]}$ R^1-R^2 R^1 = alkyl, alkynyl, aryl, vinyl R^2 = acyl, aryl, benzyl, vinyl X = Br, I, OTf, OTs
The Sonogashira Coupling	R^1————H + R^2-X $\xrightarrow[\text{base}]{\text{cat. }[Pd^0L_n]}$ R^1————R^2 R^1 = alkyl, aryl, vinyl R^3 = alkyl, benzyl, vinyl X = Br, Cl, I, OTf

The Heck Reaction	 R^4 = aryl, benzyl, vinyl X = Cl, Br, I, OTf
The Kumada Coupling	 R^4 = aryl, benzyl, vinyl X = Cl, Br, I, OTf
The Stille Coupling	R^1 = alkyl, alkynyl, aryl, vinyl R^2 = acyl, alkynyl, allyl, aryl, benzyl, vinyl X = Br, Cl, I, OAc, OP(=O)(OR)$_2$, OTf
The Suzuki Coupling	R^1 = alkyl, alkynyl, aryl, vinyl R^2 = alkyl, alkynyl, aryl, benzyl, vinyl X = Br, Cl, I, OAc, OP(=O)(OR)$_2$, OTf

These include, for example, carbohydrates and the biopolymer chitosan suitable for the Suzuki reaction [42,43] (Scheme 9). Ligands based on monosaccharides usually have only moderate solubility in water and often require incorporation of extra hydrophilic groups to improve the ligand property. Also, a co-solvent or a surfactant is sometimes needed for optimum efficiency. Dissacharide ligands have the advantage of higher aqueous solubility, but the synthesis of these is usually tedious. Chitosan, on the other hand [44-47], has proven to be a very promising material for organic synthesis and catalysis. For example, Suzuki and Heck reactions have successfully been completed using metal-chitosan complexes as catalysts in aqueous solutions [48]. The complexes have offered both economically and environmentally more favorable conditions for many reactions.

LIGANDLESS SUZUKI CROSS-COUPLING

From an environmental point of view, development of a new catalytic system without the use of stabilizing phosphine ligands in aqueous media under mild conditions has attracted much attention. The reaction with ligandless catalysts (Table 2) such as Pd/C in water has, for example, very recently been investigated [38,49].

Compared to air-sensitive and expensive homogenous palladium catalysts, palladium charcoal can safely be handled and removed from the reaction mixture by simple filtration. The recovered palladium charcoal can be purified and reused as palladium metal. These features are also of great advantage in industrial processes. Other palladium catalysts useful for the ligand less Suzuki reaction in water include $Pd(OAc)_2$ and $PdCl_2$ [50,51].

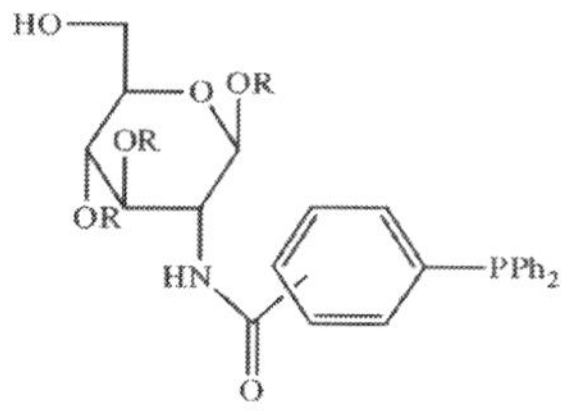

(a)

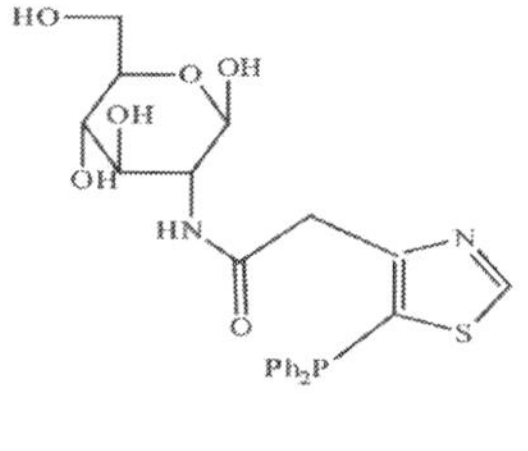

(b)

Scheme 9: Chitosan alternatives suitable for the Suzuki reaction [by 40].

TARGETS FOR THE SUZUKI CROSS-COUPLING

The synthesis of biaryl compounds, reviewed by Stanforth [55], is of importance for numerous agrochemical and pharmaceutical applications. The classic Ullmann methodology [56] is also well-known in this context. Owing, on the other hand, to the versatile chemistry of palladium compounds in carbon-carbon bond-forming reactions, several palladium-catalyzed processes have been proposed as eco-friendly replacement for this stoichiometric protocol. Essentially based on the Pd(II)-Pd(0) redox cycle, these processes require in situ regeneration of the active palladium catalyst, which can be achieved, for example, during the Suzuki reaction by various reagents (2-propanol, hydrogen gas, or aqueous alkali salts). In the synthesis of heterocyclic compounds, the aqueous Suzuki reaction has found many applications [57]. A few examples include the structural modifications and preparations of pyrroles [58], indoles [59], pyridines [60], quinoxalines [61], benzofurans [62], pyrimidyl thiazoles [63], pyridyl pyrimidines [63], and imidazoles [64]. The Suzuki reaction of boronic acids in water and a variety of heteroaryl halides has been conducted to prepare 5-substitued heteroaryl pyrimidines, which can be hydrolyzed to 5-substituted uracils as po tential antiviral agents [65]. The reaction could tolerate a broad range of functional groups, including those present in unprotected nucleotides and amino acids. In the preparation of purines on solid phase, Pd catalyses have also successfully been utilized [66]. In the synthesis of polymeric materials (Scheme 7) [24], a hydrocarbon nonmetallic conducting polymer with a rigid rod of benzene rings was synthesized recently in water [67].

CATALYTIC ASYMMETRIC SUZUKI CROSSCOUPLING

The asymmetric Suzuki cross-coupling reaction has successfully been accomplished in both organic solvents and inorganic-aqueous mixed solvents. For the preparation of C_2-symmetric biaryls, a modified Suzuki cross-coupling method of haloarenes was designed in 1996 by Keay and co-workers [68]. The catalytic asymmetric version of the reaction for the synthesis of axially chiral biaryl compounds was later developed by the Buchwald group [69].

The first, asymmetric Suzuki cross coupling performed in an organic-aqueous mixed solvent resulting in binaphthalene derivatives in up to 63% ee, was described by Cammidge and Crepy [70] (Scheme 10). In DME, the reaction gave up to 85% ee. Following these discoveries, other reports on the asymmetric synthesis for chiral binaphhyls or biaryls have been reported, for example, by Castanet et al. [71] in the study on chirality reversal depending on the palladium-chiral phosphine ratio in the reaction of sterically hindered arylboronic acids.

The reaction was performed in DME-water or toluene-EtOH-water. An atropo-enantioselective Suzuki cross coupling towards an axially chiral antimitotic biaryl has very recently been prepared by Herrbach et al. [72] in dioxane-water. Chiral atropisomeric binaphthalenes have also been synthesized in water-DME and a mixture of water-EtOH-toluene.

OPTICAL PROPERTIES OF CHOSEN SUZUKI COUPLED BRANCHED UNITS

Benzothiadiazole-based oligomers and polymers have been widely studied in recent years as active materials in various optoelectronic devices because of the heterocyclic group and the observed low band gap in polymers containing it. Copolymerization of benzothiadiazole with phenoxazine, phenothiazine, fluorene and carbazole or other suitable arylenes can be used as a means to tune the HOMO-LUMO levels in the resulting polymers. The HOMO and LUMO energy level of ϖ-conjugated polymer are important for understanding charge injection processes in the luminescent devices [24].

In flat-panel display applications the full-color capability is required, especially the blue one. Much progress has been made in the development of blue color different OLEDs devices. Many materials have been synthesized and used in blue OLEDs. Moreover, several attempts have been made to stabilize the blue colour via chemical modification, including incorporation of benzothiadiazole, anthracene and other units.

Table 2: Ligandless catalysts-examples

Type of coupling	Reaction	References
The Mizoroki – Heck coupling	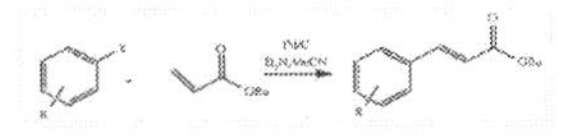	[52] Glasnov et al.
The Suzuki coupling		[53] Lu et al.
The Suzuki coupling	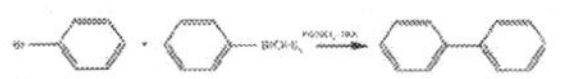	[54] Fan et al.

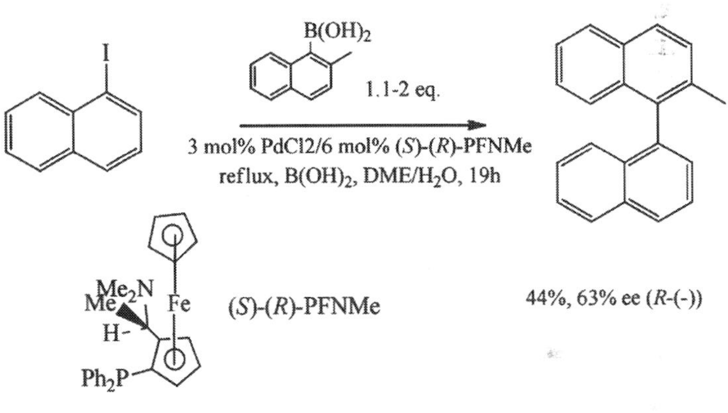

Scheme 10: Asymmetric Suzuki cross coupling [by 40]

According to the fact, it is worth to mention that the emission spectra of benzothiadiazole-based phenotiazine/phenoxazine derivatives (copolymers synthesized via Suzuki condensation) recorded at different excitation wavelengths in the range 310 - 390 nm, emitted in the blue region (Figure 2) [24]. Very similar situation we observed in case of thianthrene derivatives of phenothiazine [25].

THIOPHENE UNITS

The class of conducting polymer that has seen significant attention from both academia and industry, largely because of its diverse and often remarkable properties, is polythiophene-based polymers. The introduction of functionalized aryl moieties onto heterocyclic compounds (such as thiophenes) is an important task in organic synthesis. Three main methods exist to get unsymmetrical aryl-aryl coupling: between an aryl halide and an arylboronic acid catalysed by a palladium phosphine complex (Suzuki [8,29]), between an aryl halide and a Grignard reagent catalysed by a nickel complex (Kumada [31]), and between an aryl halide and an aryltrialkylstannane catalysed by a palladium complex (Stille [32]).

It is well established that the physical properties of conducting polymers are closely linked to the polymerization conditions and to the structure of their monomeric precursor. Polymerizations of alkylthiophenes can be classified in two categories, i.e. the oxidative electrochemical or chemical polymerization leading to positively charged polymers, and the organometallic couplings leading to neutral polymers [73]. Poly(alkylthiophenes) obtained by the first method are often high-molecular-weight polymers but with a certain amount of irregular couplings [73]. On the other hand polymers obtained by Kumada [31], Stille [32] or Suzuki [8,29] coupling reactions are more regioregular head-to-tail poly(alkylthiophenes) with high molecular weights and conductivities. The drawbacks are the need of stoichiometric amount of metal to functionalize the substrate.

The type and nature of the substituents used to induce solubility and regioregularity in oligo and polythiophenes greatly influence the morphology and band gap in these materials. Several groups have reported the synthesis of well-defined oligomers of thiophenes, especially head-totail coupled ones, to understand the influence of substitution pattern on conjugation and packing. Among the various thiophene based materials reported in the literature, regioregular poly(3-alkylthiophene)s have shown the best promise both in terms of conjugation and device performance [74]. In efforts to identify new classes of thiophene based materials with improved ϖ-conjugation and charge transport, there is i.e. reported also the synthesis and characterization of a series of structurally defined thiophene based tetramers as model compounds for polymers with thiophene based alkyl solubilizing groups. Three tetramers of thiophene (Scheme 11) bearing alkylthienyl side chains have been synthesized by palladium catalysed Suzuki cross-coupling [75].

One of the other novel thiophene structure obtained by

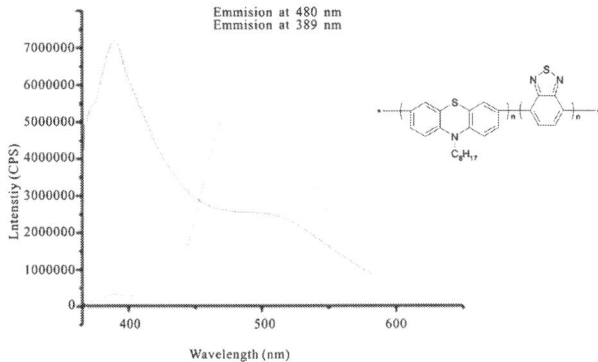

Figure 2: The fluorescence spectra of benzothiadiazole-based phenothiazine copolymer in chloroform solution (2 µM) [by 24].

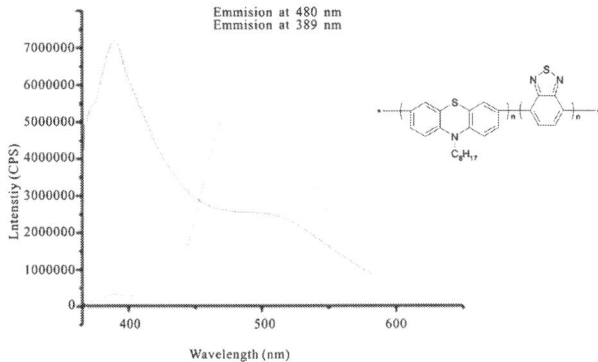

R1 = R2 = R3 = R4 =

Scheme 11: Structures of substituted thiophene-based tetramers [75].

Suzuki coupling is also the substituted thieno[3,2-b] thiophene molecules with triphenylamine core. The synthetic method allowed the preparation of trisubstituted triphenylamine derivatives of thieno[3,2-b] thiophene and thiophene (Scheme 12) in a combinatorial manner starting from tris (4-bromophenyl) amine in relatively good yields [76]. The newly suggested molecules exhibited extended ϖ-conjugation with high molar extinction coefficients and some of them demonstrated molecular glass behavior.

It is need to mention that, the pure polythiophenes are generally less suitable for polymerization by the Suzuki-Miyaura coupling method. The reason for this is the electron-rich nature of thiophenes, which slows down the oxidative addition step, resulting in an overall slower reaction and more pronounced deborylation. Deborylation is the main side reaction of the Suzuki-Miyaura coupling, resulting in low molecular weight products. For less electronegative aryl monomers like polyfluorenes, the Suzuki-Miyaura coupling is generally accepted as a better alternative than the Stille coupling, giving higher molecular weight products.

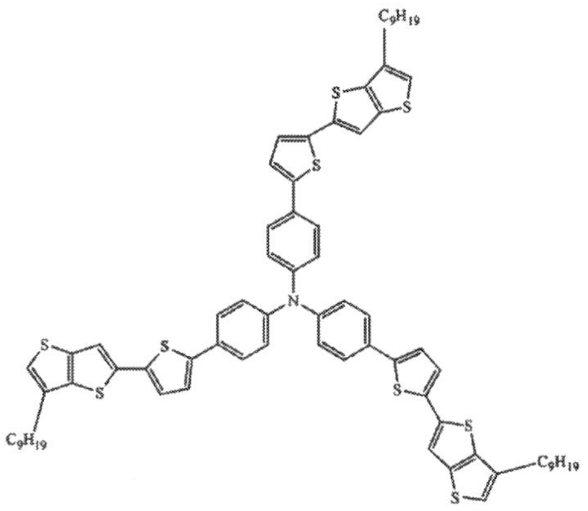

Scheme 12: Star-shaped thiophene derivative [76].

SUPRAMOLECULAR STRUCTURES BUILT OF π-STACKING THIOPHENE DERIVATIVES

The development of the methods to create nanostructures has been inspired primarily by natural environment, which displays a wide variety of complex nano-sized structures with great precision. These nano-sized structures in biological systems are specifically put together from two or more small molecular components by means of secondary interactions.

Oligoand polythiophenes form an important class of conducting materials that can find possible applications in, for example, thin film transistors or light-emitting diodes [77-79].

Previous investigations of the bulk properties of the i.e. bis(urea)-substituted thiophene derivatives under consideration have shown that the molecules form one-dimensional fibers in solution in which the thiophene moieties are ϖ-stacked [80]. Pulse-radiolysis time-resolved microwave conductivity experiments i.e. have also demonstrated that this arrangement provides an efficient path for charge transport within these self-assembled fibers. In fact, it has been shown that, by inducing a high degree of molecular order in thin films of oligothiophenes, it is possible to enhance the mobility of charge carriers [81].

The effects of the chemical composition of conjugated polymers on the supramolecular organization, i.e., the morphology of the active layer, are of major concern to further optimize the performance of the devices. This feature can sufficiently be illustrated with poly(3-alkylthiophene)s.

Since polymerization procedures for regioregular poly(3-substituted-thiophene)s have improved considerably, it is obvious that many functional features are more pronounced for the regioregular material, being either head-to-tail [82] or head-to-head/tail-to-tail, [83] due to an improved ordering of the planar conformation compared to the regioirregular polythiophene.

The latter polymer, without defined interchain interactions, gave rise to low charge-carrier mobilities (10^{-5} cm2/V·s), whereas the former, with lamellar arrays of polythiophene rods, gave significant higher mobilities (0.01 - 0.1 cm²/V·s) [84].

However, the manufacturing process of thiophene thin film may cause many problems. For instance, vacuum vapor deposition requires much energy and expensive equipment. Although thin film fabrication method via simple wet process have been developed using a polymer solution, it is difficult to obtain polymer thin films with high crystallinity. Ikeda and co-worker overcame these problems and found a facile manufacturing method of thiophene nanosheets with high crystallinity in the solution [85]. They discovered that an alternating copolymer, in which a thiophene derivative and flexible ethylene glycol chain are alternately connected, is folded in some organic solvents in such a way that the thiophene units are stacked each other, and the folded copolymers self-assemble into a 2-dimensional sheet structure (Scheme 13).

The arrangement of the thiophene units in the nanosheet was confirmed to be the same as that manufactured by vacuum vapor-deposition of low-molecularweight thiophene compounds. Therefore, our thiophene nanosheets are feasible to the application of organic electronics devices [85].

These tailor-made supramolecular assemblies will enormously influence the macroscopic properties. The combination of supramolecular architecture with functionality in macromolecules will not only give rise to emerging opportunities in materials science, but also significantly contribute to bridge the gap between natural and artificial systems in an effort to fully understand the guidelines used to assemble natural units in the different hierarchies of organization.

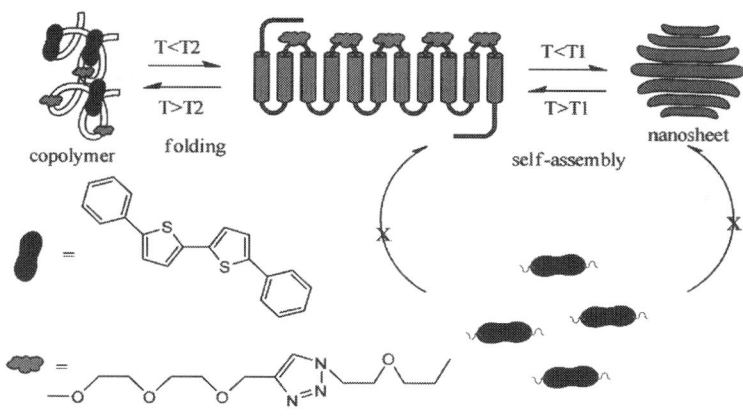

Scheme 13: Concept of Ikada et al. of the supramolecular thiophene nanosheet formation.

CONCLUDING REMARKS

Carbon-carbon bond formations reactions on solid-support are highly useful and versatile methods needed for the development of modern organic synthesis discovery.

The review focuses on the investigation of the solidphase adaption of the Suzuki, the coupling reactions due the fact that these have proven to be both reliable and efficient methods.

Although the method has found many applications in synthesizing different molecules, there is still much work on the development towards an efficient catalyst applicable for structurally different substrates. The work on developing a catalyst working entirely in water has resulted in a few remarkable discoveries. However, during the next few years, it is anticipated that new and more efficient catalysts will be discovered, giving even more excellent results in aqueous media.

Since the first reports by Miyaura and Suzuki in 1979, the palladium-catalyzed cross-coupling reactions of organoboron derivatives with organic electrophiles have been widely employed in both academic and industrial laboratories and have significantly contributed to the efficient synthesis also on a large scale of a wide range of interesting compounds including pharmaceuticals, agrochemicals and synthetic intermediates.

In spite of the fact that, in recent years, many new transition metal-catalyzed methods allowing C-C bond formation have been developed, the Suzuki-Miyaura reaction is still one of the most reliable and environmentally friendly tools for installing a wide range of nonfunctionalized and functionalized carbon substituents on (hetero)aromatic systems with exquisite chemoand site-selectivity and is by far the most versatile method for the synthesis of industrially relevant functionalized heteroarenes and unsymmetrical biheteroaryl derivatives.

Finally, summarizing the results illustrated in this paper, it may be noted that for substrates bearing different halogen atoms chemoselective cross-couplings have been accomplished on the basis of the reactivity difference between the halogens. However, the hardest acquirable selectivity in Suzuki-Miyaura monocoupling reactions involving polyhalogenated heteroarenes bearing identical halogen atoms has been shown to be dominated by steric and electronic effects and the presence of directing groups at positions neighboring the reaction sites. Moreover, for symmetrically substituted dihalogenated heteroarenes bearing identical halogen atoms its achievement has sometimes required a careful optimization of the reaction parameters including the nature of the catalyst precursor, the base, the solvent and the molar ratio between electrophile and organoboron reagent.

Interestingly, the efficiency and selectivity of several monoarylation reactions illustrated in the sections of this review have frequently allowed the development of one-pot polycoupling procedures, but, unfortunately, this synthetically useful approach to polysubstituted heteroarenes has not been applied so far to the synthesis of important pharmacologically active agents and structurally complex natural products.

It is believed that many of the one-pot processes developed so far, which are valuable for laboratory syntheses on a small scale, require further studies aimed to make them suitable for large-scale preparations through improvement of their yields and selectivity and development of more efficient and cost effective catalyst systems.

ACKNOWLEDGEMENTS

The authors are gratefully acknowledged for the financial support of the Polish National Centre of Progress of Explorations Grant no. NR05-0017-10/2010 and Wroclaw University of Technology.

REFERENCES

1.   R. Franzen, "The Suzuki, the Heck, and the Stille Reaction/Three Versatile Methods for the Introduction of New C-C Bonds on Solid Support," Canadian Journal of Chemistry, Vol. 78, No. 7, 2000, pp. 957-962.

2.   J. M. Davidson and C. J. Triggs, "Reaction of Metal Ion Complexes with Hydrocarbons. I. Palladation and Some Other New Electrophilic Substitution Reactions. Preparation of Palladium(I)," Journal of the Chemical Society A: Inorganic, Physical, Theoretical, Vol. 6, 1968, pp. 1324-1331.

3.   K. Garves, "Coupling, Carbonylation, and Vinylation Reactions of Aromatic Sulfinic Acids via Organopalladium Intermediates," The Journal of Organic Chemistry, Vol. 35, No. 10, 1970, pp. 3273-3275.

4.   N. Miyaura and A. Suzuki, "Stereoselective Synthesis of Arylated (E)-Alkenes by the Reaction of Alk-1-enylboranes with Aryl Halides in the Presence of Palladium Catalyst," Journal of the Chemical Society, Chemical Communications, Vol. 19, 1979, pp. 866-867.

5. A. Nowakowska-Oleksy, J. Sołoducho and J. Cabaj, "Phenoxazine Based Units—Synthesis, Photophysics and Electrochemistry," Journal of Fluorescence, Vol. 21, No. 1, 2011, pp. 169-178.

6. J. Sołoducho, J. Cabaj, K. Idzik, A. Nowakowska-Oleksy, A. wist and M. Łapkowski, "Synthesis, Structure and Properties of Crowded Symmetric Arylenes," Current Organic Chemistry, Vol. 14, No. 12, 2010, pp. 1234-1244.

7. N. Miyaura, K. Yamada and A. Suzuki, "A New Stereospecific Cross-Coupling by the Palladium-Catalyzed Reaction of 1-Alkenylboranes with 1-Alkenyl or 1-Alkynyl Halides," Tetrahedron Letters, Vol. 20, No. 36, 1979, pp. 3437-3440.

8. N. Miyaura and A. Suzuki, "Palladium-Catalyzed CrossCoupling Reactions of Organoboron Compounds," Chemical Reviews, Vol. 95, No. 7, 1995, pp. 2457-2483.

9. G. Altenhoff, R. Goddard, C. W. Lehmann and F. Glorius, "A N-Heterocyclic Carbene Ligand with Flexible Steric Bulk Allows Suzuki Cross-Coupling of Sterically Hindered Aryl Chlorides at Room Temperature," Angewandte Chemie International Edition, Vol. 42, No. 31, 2003, pp. 3690-3693.

10. W. A. Herrmann, K. Fele, S. K. Schneider, E. Herdtweck and S. D. Hoffmann, "A Carbocyclic Carbene as an Efficient Catalyst Ligand for C-C Coupling Reactions," Angewandte Chemie International Edition, Vol. 45, No. 23, 2006, pp. 3859-3862.

11. W. J. Tang, A. G. Capacci, X. D. Wei, W. J. Li, A. White, N. D. Patel, J. Savoie, J. J. Gao, S. Rodriguez, B. Qu, N. Haddad, B. Z. Lu, D. Krishnamurthy, N. K. Yee and C. H. Senanayake, "A General and Special Catalyst for SuzukiMiyaura Coupling Processes," Angewandte Chemie, Vol. 122, No. 34, 2010, pp. 6015-6019.

12. D. H. Lee and M. J. Jin, "An Extremely Active and General Catalyst for Suzuki Coupling Reaction of Unreactive Aryl Chlorides" Organic Letters, Vol. 13, No. 2, 2011, pp. 252-255.

13. M. Thimmaiah and S. Fang, "Efficient Palladium-Catalyzed Suzuki-Miyaura Coupling of Aryl Chlorides with Arylboronic Acids Using Benzoferrocenyl Phosphines as Supporting Ligands," Tetrahedron, Vol. 63, No. 29, 2007, pp. 6879-6886.

14. Y. Uozumi, Y. Matsuura, T. Arakawa and Y. M. A. Yamada, "Asymmetric Suzuki-Miyaura Coupling in Water with a Chiral Palladium Catalyst Supported on an Amphiphilic Resin"

Angewandte Chemie International Edition, Vol. 48, No. 15, 2009, pp. 2708-2710.

15. D. J. M. Snelders, G. van Koten and R. Gebbink, "Hexacationic Dendriphos Ligands in the Pd-Catalyzed Suzuki-Miyaura Cross-Coupling Reaction: Scope and Mechanistic Studies," Journal of the American Chemical Society, Vol. 131, No. 32, 2009, pp. 11407-11416.

16. L. Yin and J. Liebscher, "Carbon-Carbon Coupling Reactions Catalyzed by Heterogeneous Palladium Catalysts," Chemical Reviews, Vol. 107, No. 1, 2006, pp. 133-173.

17. M. Lysen and K. Kohler, "Suzuki-Miyaura Cross-Coupling of Aryl Chlorides in Water Using Ligandless Palladium on Activated Carbon," Synlett, Vol. 11, 2005, pp. 1671-1674.

18. M. Lysen and K. Kohler, "Palladium on Activated Carbon—A Recyclable Catalyst for Suzuki-Miyaura CrossCoupling of Aryl Chlorides in Water," Synthesis, Vol. 4, 2006, pp. 692-698.

19. N. Gürbüz, I. Özdemir, T. Seçkin, B. Çetinkaya, "Surface Modification of Inorganic Oxide Particles with a Carbene Complex of Palladium: A Recyclable Catalyst for the Suzuki Reaction," Journal of Inorganic and Organometallic Polymers, Vol. 14, 2No. 2, 004, pp.149-159.

20. J. Han, Y. Liu and R. Guo, "Facile Synthesis of Highly Stable Gold Nanoparticles and Their Unexpected Excellent Catalytic Activity for Suzuki-Miyaura Cross-Coupling Reaction in Water," Journal of the American Chemical Society, Vol. 131, No. 6, 2009, pp. 2060-2061.

21. B. Z. Yuan, Y. Y. Pan, Y. W. Li, B. L. Yin and H. F. Jiang, "A Highly Active Heterogeneous Palladium Catalyst for the Suzuki-Miyaura and Ullmann Coupling Reactions of Aryl Chlorides in Aqueous Media," Angewandte Chemie International Edition, Vol. 49, No. 24, 2010, pp. 4054-4058.

22. M. J. Jin and D. H. Lee, "A practical Heterogeneous Catalyst for the Suzuki, Sonogashira, and Stille Coupling Reactions of Unreactive Aryl Chlorides," Angewandte Chemie International Edition, Vol. 49, No. 6, 2010, pp. 1119-1122.

23. B. Li, Z. Guan, W. Wang, X. Yang, J. Hu, B. Tan and T. Li, "Highly Dispersed pd Catalyst Locked in Knitting Aryl Network Polymers

for Suzuki-Miyaura Coupling Reactions of Aryl Chlorides in Aqueous Media," Advanced Materials, Vol. 24, 2No. 25, 012, pp. 3390-3395.

24. A. Nowakowska-Oleksy, J. Cabaj, K. Olech and J. Sołoducho, "Comparative Study of Alternating Low-BandGap Benzothiadiazole Co-Oligomers," Journal of Fluorescence, Vol. 21, No. 4, 2011, pp. 1625-1633.

25. A. wist, J. Sołoducho, P. Data and M. Łapkowski, "Thianthrene-Based Oligomers as Hole Transporting Materials," ARKIVOC, Vol. 2012, No. 3, 2012, pp. 193-209.

26. M. Sailer, R. A. Gropeanu and T. J. Müller, "Practical Synthesis of Iodo Phenothiazines. A Facile Access to Electrophore Building Blocks," The Journal of Organic Chemistry, Vol. 68, No. 19, 2003, pp. 7509-7512. doi:10.1021/jo034555z

27. P. Herguth, X. Jiang, M. S. Liu and A. K. Y. Jen, "Highly Efficient Fluoreneand Benzothiadiazole-Based Conjugated Copolymers for Polymer Light-Emitting Diodes," Macromolecules, Vol. 35, No. 16, 2002, pp. 6094-6100.

28. Y. Zhu, R. D. Champion and S. A. Jenekhe, "Conjugated Donor-Acceptor Copolymer Semiconductors with Larg Intramolecular Charge Transfer: Synthesis, Optical Properties, Electrochemistry, and Field Effect Carrier Mobility of Thienopyrazine Based Copolymers," Macromolecules, Vol. 39, No. 25, 2006, pp. 8712-8719.

29. N. Miyaura and A. Suzuki, "Stereoselective Synthesis of Arylated (E)-Alkenes by the Reaction of Alk-1-Enylboranes with Aryl Halides in the Presence of Palladium Catalyst," Journal of the Chemical Society, Chemical Communications, Vol. 19, 1979, pp. 866-867.

30. R. F. Heck, "Palladium-Catalysed Vinylation of Organic Halides," Organic Reactions, Vol. 27, 1982, pp. 345-390.

31. T. Hayashi and M. Kumada, "Asymmetric Synthesis Catalyzed by Transition-Metal Complexes with Functionalized Chiral Ferrocenylphosphine Ligands," Accounts of Chemical Research, Vol. 15, No. 12, 1982, pp. 395-401. doi:10.1021/ar00084a003

32. J. W. Labadie, D. Tueting and S. K. Stille, "Synthetic Utility of the Palladium-Catalyzed Coupling Reaction of Acid Chlorides with

Organotins," The Journal of Organic Chemistry, Vol. 48, No. 24, 1983, pp. 4634-4642. doi:10.1021/jo00172a038

33. E. Negishi, "Palladiumor Nickel-Catalyzed cross Coupling. A New Selective Method for Carbon-Carbon Bond Formation," Accounts of Chemical Research, Vol. 15, No. 11, 1982, pp. 340-348. doi:10.1021/ar00083a001

34. K. Sonogashira, "Development of Pd-Cu Catalyzed Cross-Coupling of Terminal Acetylenes with sp^2-Carbon Halides," Journal of Organometallic Chemistry, Vol. 653, No. 1-2, 2002, pp. 46-49. doi:10.1016/S0022-328X(02)01158-0

35. J. Tsuji, "Transition metal reagents and catalysts: Innovations in organic synthesis," John Wiley and Sons, New York, 2002. doi:10.1002/0470854766M. F. Lipton, M. A. Mauragis, M. T. Maloney, M. F. Veley, D. W. Vander-Bor, J. J. Newby, R. B. Appell and E. D. Daugs, "The Synthesis of OSU 6162: Efficient, Large-Scale Implementation of a Suzuki Coupling," Organic Process Research and Development, Vol. 7, No. 3, 2003, pp. 385-392. doi:10.1021/op025620u

36. M. Beller and A. Zapf, "Handbook of Organopalladium Chemistry for Organic Synthesis," Wiley, Hoboken, 2002.

37. D. S. Ennis, J. McManus, W. Wood-Kaczmar, J. Richardson, G. E. Smith and A. Carstairs, "Multikilogram-Scale Synthesis at a Biphenyl Carboxylic Acid Derivative Using a Pd/C-Mediated Suzuki Coupling Approach," Organic Process Research and Development, Vol. 3, No. 4, 1999, pp. 248-252. doi:10.1021/op980079g

38. Y. Mori, M. Nakamura, T. Wakabayashi, K. Mori and S. Kobayashi, "Efficient Total Synthesis of Khafrefungin: Convergent Approach Using Suzuki Coupling under Thallium-Free Conditions toward Multigram-Scale Synthesis," Synlett, Vol. 4, 2002, pp. 601-603. doi:10.1055/s-2002-22726

39. R. Franzén and Y. Xu, "Review on Green Chemistry —Suzuki Cross-Coupling in Aqueous Media," Canadian Journal of Chemistry, Vol. 83, No. 3, 2005, pp. 266-272.doi:10.1139/v05-048

40. A. Herrmann and C. W. Kohlpaintner, "Water-Soluble Ligands, Metal Complexes, and Catalysts: Synergism of Homogeneous and Heterogeneous Catalysis," Angewandte Chemie International

Edition in English, Vol. 32, No. 11, 1993, pp. 1524-1544. doi:10.1002/anie.199315241

41. U. M. Lindström, "Stereoselective Organic Reactions in Water," Chemical Reviews, Vol. 102, No. 8, 2002, pp. 2751-2772.

42. M. Bellar, J. G. E. Krauter and A. Zapf, "Kohlenhydrat-substituierte Triarylphosphane-Eine Neue Ligandenklasse für die Zweiphasen-Katalyse," Angewandte Chemie International Edition in English, Vol. 36, No. , 1997, pp. 772-774.

43. J. Zhang and C.-G. Xia, "Kinetic Study of Dichlorocyclopropanation of 4-Vinyl-1-cyclohexene by a Novel Multisite Phase Transfer Catalyst," Journal of Molecular Catalysis A: Chemical, Vol. 206, No. 1-2, 2003, pp. 59-68.

44. T. Vincent and E. Guibal, "Chitosan-Supported Palladium Catalyst. Synthesis Procedure," Industrial and Engineering Chemistry Research, Vol. 41, No. 21, 2002, pp. 5158- 5164.doi:10.1021/ie0201462

45. F. Quignard, A. Choplin and A. Damard, "Chitosan: A Natural Polymeric Support of Catalysts for the Synthesis of Fine Chemicals," Langmuir, Vol. 16, No. 24, 2000, pp. 9106-9108. doi:10.1021/la000937d

46. M.-A. Yin, G.-L. Yuan, Y.-Q. Wu, M.-Y. Huang and Y.-Y. Jiang, "Asymmetric Hydrogenation of Ketones Catalyzed by a Silica-Supported Chitosan-Palladium Complex," Journal of Molecular Catalysis A: Chemical, Vol. 147, No. 1-2, 1999, pp. 93-98. doi:10.1016/S1381-1169(99)00133-8

47. J. J. E. Hardy, S. Hubert, D. C. Macquarrie and A. J. Wilson, "Chitosan-Based Palladium Catalysts in the Heck and Suzuki Reactions," Green Chemistry, Vol. 6, 2004, p. 53.doi:10.1039/b312145n

48. H. Sakurai, T. Tsukuda and T. Hirao, "Pd/C as a Reusable Catalyst for the Coupling Reaction of Halophenols and Arylboronic Acids in Aqueous Media," Journal of Organic Chemistry, Vol. 67, No. 8, 2002, pp. 2721-2722. doi:10.1021/jo016342k

49. M. T. Reetz and E. Westermann, "Phosphanfreie Palladium-katalysierte Kupplungen: Die Entscheidende Rolle von Pd-Nanoteilchen," Angewandte Chemie, Vol. 112, No. 1, 2000, pp. 170-173. doi:10.1002/(SICI)1521-3757(20000103)112:1<170::AID-ANGE170>3.0.CO;2-A

50. G. W. Kabalka, V. Namboodiri and L. Wang, "Suzuki Coupling with Ligandless Palladium and Potassium Fluoride," Chemical Communications, Vol. 8, 2001, p. 775.doi:10.1039/b101470f

51. T. N. Glasnov, S. Findening and C. O. Kappe, "Heterogeneous versus Homogeneous Palladium Catalysts for Ligandless Mizoroki-Heck Reactions: A Comparison of Batch/Microwave and Continuous-Flow Processing," Chemistry—A European Journal, Vol. 15, No. 4, 2009, pp. 1001-1010. doi:10.1002/chem.200802200

52. G. Lu, R. Franzen, Q. Zhang and Y. Xu, "Palladium Charcoal-Catalyzed, Ligandless Suzuki Reaction by Using Tetraarylborates in Water," Tetrahedron Letters, Vol. 46, No. 24, 2005, pp. 4255-4259. doi:10.1016/j.tetlet.2005.04.022

53. G. Fan, B. Zou, S. Cheng and L. Zheng, "Ligandless Palladium Supported on SiO_2-TiO_2 as Effective Catalyst for Suzuki Reaction," Journal of Industrial and Engineering Chemistry, Vol. 16, No. 2, 2010, pp. 220-223. doi:10.1016/j.jiec.2009.08.009

54. S. P. Stanforth, "Catalytic Cross-Coupling Reactions in Biaryl Synthesis," Tetrahedron, Vol. 54, No. 3-4, 1998, pp. 263-303. doi:10.1016/S0040-4020(97)10233-2

55. F. Ullmann, "Ueber eine neue Darstellungsweise von Phenyläthersalicylsäure," Berichte der Deutschen Chemischen Gesellschaft, Vol. 37, No. 1, 1904, pp. 853-854.

56. J. J. Li and G. W. Gribble, "Palladium in Heterocyclic Chemistry," Pergamon, New York, 2000.

57. C. N. Johnson, G. Stemp, N. Anand, S. C. Stephen and T. Gallagher, "Palladium(0)-catalysed Arylations Using Pyrrole and Indole-2-boronic Acids," Synlett, No. 9, 1998, pp. 1025-1027. doi:10.1055/s-1998-1834

58. R. J. Sundberg, "Indoles," Academic Press, London, 1996.

59. A. Zoltewicz and M. P. Cruskie Jr., "Strategies for the Synthesis of Unsymmetrical Quaterpyridines Using Palladium-Catalyzed Cross-Coupling Reactions," Tetrahedron, Vol. 51, No. 42, 1995, pp. 11393-11400. doi:10.1016/0040-4020(95)00699-9

60. J. Li and W. S. Yue, "Synthesis of 3-Aryl and 3-Heterocyclic Quinoxalin-2-ylamines via Pd-Catalyzed CrossCoupling

Reactions," Tetrahedron Letters, Vol. 40, No. 24, 1999, pp. 4507-4510. doi:10.1016/S0040-4039(99)00822-9

61. T. Soos, G. Timari and G. Hajos, "A New and Concise Synthesis of Furostifoline," Tetrahedron Letters, Vol. 40, No. 49, 1999, pp. 8607-8609. doi:10.1016/S0040-4039(99)01803-1

62. S. L. Hargreaves, B. L. Pilkington, S. E. Russell and P. A. Worthington, "The Synthesis of Substituted Pyridylpyrimidine Fungicides Using Palladium-Catalysed CrossCoupling Reactions," Tetrahedron Letters, Vol. 41, No. 10, 2000, pp. 1653-1656. doi:10.1016/S0040-4039(00)00007-1

63. D. Wang and J. Haseltine, "A Comparison of Phenylboronic Acid and Phenyltrimethyltin in the Palladium-Catalyzed Arylation of 1,5-Dialkylimidazoles," Journal of Heterocyclic Chemistry, Vol. 31, No. 6, 1994, pp. 1637-1639. doi:10.1002/jhet.5570310660

64. D. Peters, A.-B. Hörnfeldt and S. Gronovitz, "Synthesis of Various 5-Substituted Uracils," Journal of Heterocyclic Chemistry, Vol. 27, No. 7, 1990, pp. 2165-2173.doi:10.1002/jhet.5570270756

65. R. Franzén and J. Tois, "Purine and Sugar Chemistry on Solid Phase—100 Years after the Emil Fischer's Chemistry Nobel Prize 1902," Combinatorial Chemistry & High Throughput Screening, Vol. 6, No. 5, 2003, pp. 433-434. doi:10.2174/138620703106298617

66. T. I. Wallow and B. M. Novak, "In Aqua Synthesis of Water-Soluble Poly(para-phenylene) Derivatives," Journal of the American Chemical Society, Vol. 113, No. 19, 1991, pp. 7411-7412. doi:10.1021/ja00019a042

67. N. G. Andersen, S. P. Maddaford and B. A. Keay, "A Modified in Situ Suzuki Cross-Coupling of Haloarenes for the Preparation of C_2 Symmetric Biaryls," Journal of Organic Chemistry, Vol. 61, No. 26, 1996, pp. 9556-9559. doi:10.1021/jo9617880

68. J. Yin and S. L. Buchwald, "A Catalytic Asymmetric Suzuki Coupling for the Synthesis of Axially Chiral Biaryl Compounds," Journal of the American Chemical Society, Vol. 122, No. 48, 2000, pp. 12051-12052. doi:10.1021/ja005622z

69. A. N. Cammidge and K. V. L. Crepy, "The First Asymmetric Suzuki Reaction," Chemical Communications, Vol. 18, 2000, pp. 1723-1724. doi:10.1039/b004513f

70. A.-S. Castanet, F. Colobert, P.-E. Broutin and M. Obringer, "Asymmetric Suzuki Cross-Coupling Reaction: Chirality Reversal Depending on the Palladium-Chiral Phosphine Ratio," Tetrahedron: Asymmetry, Vol. 13, No. 6, 2002, pp. 659-665. doi:10.1016/S0957-4166(02)00169-6

71. A. Herrbach, A. Marinetti, O. Baudoin, D. Guenard and F. Gueritte, "Asymmetric Synthesis of an Axially Chiral Antimitotic Biaryl via an Atropo-Enantioselective Suzuki Cross-Coupling," Journal of Organic Chemistry, Vol. 68, No. 12, 2003, pp. 4897-4905.doi:10.1021/jo034298y

72. M. Sevignon, J. Hassan, C. Gozzi, E. Schulz and M. Lemaire, "A New Green Catalytic Method for Biaryl CrossCoupling and Oligothiophene Synthesis," Comptes Rendus de l'Académie des Sciences-Series IIC-Chemistry, Vol. 3, No. 7, 2000, pp. 569-572.

73. I. Osaka and R. D. McCullough, "Advances in Molecular Design and Synthesis of Regioregular Polythiophenes," Accounts of Chemical Research, Vol. 41, No. 9, 2008, pp. 1202-1214. doi:10.1021/ar800130s

74. G. Saini, N. T. Lucas and J. Jacob, "Tetrathiophenes with Thiophene Side Chains: Effect of Substitution on Packing and Conjugation," Tetrahedron Letters, Vol. 51, No. 22, 2010, pp. 2956-2958. doi:10.1016/j.tetlet.2010.03.087

75. N. Metri, X. Sallenave, L. Beouch, C. Plesse, F. Goubard and C. Chevrot, "New Star-Shaped Molecules Derived from Thieno[3,2-b]thiophene Unit and Triphenylamine," Tetrahedron Letters, Vol. 51, No. 50, 2010, pp. 6673-6676.doi:10.1016/j. tetlet.2010.10.082

76. R. H. Friend, R. W. Gymer, A. B. Holmes, J. H. Burroughes, R. N. Marks, C. Taliani, D. D. C. Bradley, D. A. dos Santos, M. Logdlund and W. R. Salaneck, "Electroluminescence in Conjugated Polymers," Nature, Vol. 397, 1999, pp. 121-128. doi:10.1038/16393

77. D. Fichou, "Handbook of Oligoand Polythiophene," Wiley-VCH, Weinheim, 1998.doi:10.1002/9783527611713

78. Z. Bao, A. Lovinger and J. Brown, "New Air-Stable n-Channel Organic Thin Film Transistors," Journal of the American Chemistry Society, Vol. 120, No. 1, 1998, pp. 207-208. doi:10.1021/ ja9727629

79. D. B. A. Rep, R. Roelfsema, J. H. van Esch, F. S. Schoonbeek, R. M. Kellogg, B. L. Feringa, T. T. M. Palstra and T. M. Klapwijk, "Self-Assembly of Low-Dimensional Arrays of Thiophene Oligomers from Solution on Solid Substrates," Advanced Materials, Vol. 12, No. 8, 2000, pp. 563-566. doi:10.1002/(SICI)1521-4095(200004)12:8<563::AID-ADMA563>3.0.CO;2-7

80. F. Garnier, A. Yasser, R. Hajlaoui, G. Horowitz, F. Deloffre, B. Servet, S. Ries and P. Alnot, "Molecular Engineering of Organic Semiconductors: Design of Self-Assembly Properties in Conjugated Thiophene Oligomers," Journal of the American Chemistry Society, Vol. 115, No. 19, 1993, pp. 8716-8721. doi:10.1021/ja00072a026

81. R. D. McCullough, R. D. Lowe, M. Jayaraman, P. C. Ewbank, D. L. Anderson and S. Tristram-Nagle, "Novel Coordination Complexes of Tetrathiafulvalene Derivatives," Synthetic Metals, Vol. 55, No. 1, 1993, pp. 1198-1203.

82. K. Faid, R. Cloutier and M. Leclerc, "Design of Novel Electroactive Polybithiophene Derivatives," Macromolecules, Vol. 26, No. 10, 1993, pp. 2501-2507. doi:10.1021/ma00062a017

83. H. Sirringhaus, P. J. Brown, R. H. Friend, M. M. Nielsen, K. Bechgaard, B. M. W. Langeveld-Voss, A. J. H. Spiering, R. A. J. Janssen, E. W. Meijer, P. Herwig and D. M. De Leeuw, "Two-Dimensional Charge Transport in SelfOrganized, High-Mobility Conjugated Polymers," Nature, Vol. 401, 1999, pp. 685-688. doi:10.1038/44359

84. Y. Zheng, H. Zhou, D. Liu, G. Floudas, M. Wagner, K. Koynov, M. Mezger, H. J. Butt and T. Ikeda, "Supramolecular Thiophene Nanosheets," Angewandte Chemie International Edition, Vol. 52, No. 18, 2013, pp. 4845-4848. doi:10.1002/anie.201210090

Experimental Study of Drying Process of COLZA Seeds in Fluidized Bed Dryer by Statistical Methods

Jamshid Khorshidi[1] and Hassan Davari[2]

[1]Department of Mechanical Engineering, Hormozgan University, Bandar Abbas, Iran

[2]Department of Mechanical Engineering, Islamic Azad University, Roudan Branch, Roudan, Iran

ABSTRACT

In this study the effect of initial parameters such as inlet gas temperature, initial particles temperature and gas velocity on temperature changes of solid particles and outlet gas temperature in a fluidized bed dryer was studied. For testing, an experimental setup was established. With combination of air and Colza seeds belonging to D groups of the Geldart classification (Geldart, 1986) fluidization regime was carried out. With five test series with maintaining the inlet gas temperature, solid particle temperature and outlet gas temperature during time were carefully measured. To analyze these data by using regression analysis to predict solid particle and outlet gas temperature, 2 correlations on initial parameters were presented. The result has shown that temperature gradients in the beginning of fluidization, is very high and therefore the exponential functions in the regression model is used to predict the temperature changes.

INTRODUCTION

Fluidization is the phenomenon in which solid particles in a gas or a liquid type are suspended and it has many applications in many physical, chemical industries. One of the most prevalent implementation of this phenomenon is to dry granular seed. Fluidized bed dryers have many usages in chemical, agricultural and medical industries. The Major reason to use such dryers in those industries is:

- Height heat and mass transfer coefficients due to gas-solid contact;
- High quality in produced products because of harmony and solid-gas proper mixture;
- They are suitable for operations in great scale;
- They have low service and maintenance cost;
- Gas flow voids particles crack and fraction.

High application of such dryers has led to many researches in this field which mostly are depended an experimental equations and today many experimental equations are available to predict heat and mass transfer coefficients provided from these researches. The significant point is that in all of these researches, every equation has been presented in specific condition limit of fluidization regime type so conditions dominated on problem have significant importance to use these equations because every equation has validity on specific domain of particles type, fluidization regime, specific pressure and temperature.

Geldart categorized for the time fluidization regimes of solid-gas in to 4 groups A, B, C, and D by running precise tests [1]. These classifications are based on density, solid particle diameters, gas density. In Figure 1 you can see related diagram about this classification. Botteril et al. analyzed pressure effect on heat transfer coefficient between bed and a suspended surface [2]. The results showed that temperance transference coefficient between a bed and a suspended surface increased as bed pressure enhanced. Also results proved that pressure effect on heat transfer coefficient reduced as particle size is decreased.

Hariprasad et al. evaluated temperature effect on minimum fluidization velocity (U_{mf}) in their study [3]. They ran the tests for 9 kinds of different particles belonged to group B of Geldart classification at thermal range of 298 to 973 Kelvin and calculated minimum fluidization velocity and presented equations for U_{mf}. In performed tests, experimental data belonged to U_{mf} compared to other equations which were obtained by other researchers.

Rizzi et al. [4] used a laboratory device to evaluate heat transfer in a fluidized bed containing grass seeds which belonged to group D of Geldart classification and finally presented an equation to predict heat transfer coefficient based on Reyrold number. Following to Khorshidi et al. reformed modeling and applied more proper equations and used the same data to analyze heat transference phenomenon in such dryers [5].

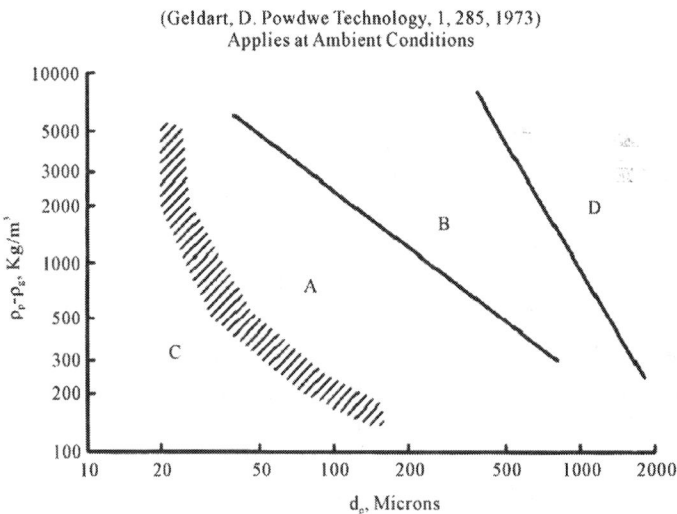

(Geldart, D. Powdwe Technology, 1, 285, 1973)
Applies at Ambient Conditions

A: Aeratable ($U_{mb} > U_{mf}$) Material Has a Significant Deaeration Time (*FCC Catalyst*)
B: Bubbles Above U_{mf} ($U_{mb} = U_{mf}$) (*500-micron Sand*)
C: Cohesive (*Flour, Fly Ash*)
D: Spoutable (*Wheat, 2000-micron Polyethylene Pellets*)

Figure 1: Geldart classification diagram.

Our purpose in this research is to evaluate variations of temperature in one fluidized bed dryer contains Colza seeds belonged to group D of Geldart classification by experimental tests in order to present temperature variation which Manifests high rate of heat transfer in such dryer.

To analyze provides data we applied statistical method based on Regression model which is used to point out to studies related to variables relations and it was expanded for the first time by Karl Pearson for statistical context.

MATERIALS AND METHODS

Particle Characterization

The consumed seed for these tests was Colza seed. Oily seeds are the secondary consumable global sources after cereals. Colza or Canola is the scientific name for Brassica Napus which is the third oily plant in the world Colza seed contain 40% to 45% oil and 30% to 35% proteins and because of so it is considered as one of the most important oily seed in the world. Colza oil mainly used in nutritive, color, chemical, lubrication, Soap, knitting and leather industries. In 1957 in Canada, the first oily Colza regenerated by a few amount of Oursic acid. To produce Colza as much as possible during 1965, thousands hectares of Canada fields were used specifically to till this plant. In 1971 span type, the first variated with low oursic acid and three years later after tower type with little Oursize acid and Glucoseinolat as the first Kahola variations were introduced. Kanola some was registered in 1978 by Canadian oil extracting institute. Several methods are used to dry such oily seeds but Gazor et al. (2008) proved in their study that fluidization method for this grain drying has had developmental effect on some of quality specifications of extracted oil like color and acidity in addition to drying time meaningful reduction.

Experimental Setup

For experimental modeling in this research, an experimental setup which makes a fluidized bed was established, this device contains a compressor with 2 hp power capability that inters air with controlled pressure and flow in to an electric heater by 1000 Watt power and then send it in to a cylindrical chamber in which there is fluidized bed. This system has controllers for pressure and temperature to measure and control them in different sections. This cylindrical bed diameter is 3 cm with external diameter of 3.5 cm made of glass with two holes that one of them is located to install solid particles temperature sensor at 2 cm stature and second holes is located at 12 cm height for outlet gas temperature sensor to receive data from the bed. By 4 thermocouple type K with 0.1 accuracy on centigrade, the electronic heater temperature, and inlet gas temperature, solid particles and outlet gas temperatures are measured. To control inlet gas temperature to the bed a PID[1] controller is used. To prevent thermal loss, fiber glass has been used as thermal isolation of bed walls. We use a distributor plate in the bed to make a uniform gas flow. This plate has holes reticulatedly with 1 mm diameter and 1 cm thickness. To weight solid particles a digital balance with 0.01 gram precision (model Scont Pro Spu 902) has been used. In following picture you can see schematic (Figure 2) of this device:

The physical properties of Colza seeds and the Geometric properties of the bed used in the experiments are shown in Table 1.

In Experimental modeling, solid particles temperature (T_s) and outlet gas temperature from bed (T_{gl}) during 20 minutes have been registered by keeping inlet gas temperature steady and fixed. Also we measure velocity and fluidization high and repeat tests for different inlet gas temperatures and velocity for 5 conditions and then registered the results should be pointed that all experimental data are provided in bubbling regime. In Table 2 you can see operational conditions of experiments.

Data related to solid particle temperature and outlet gas temperature from bed are obtained in a condition at which inlet gas from bed are obtained in a condition where inlet gas temperature is kept stable by controller system by 0.1 precision and because of bed isolation, there is no thermal losses from bed zone.

RESULTS AND DISCUSSION

Descriptive Statistics

As Table 3 shows solid particles mean temperature is 42.43 and out let gas mean temperature is 40.95 and in

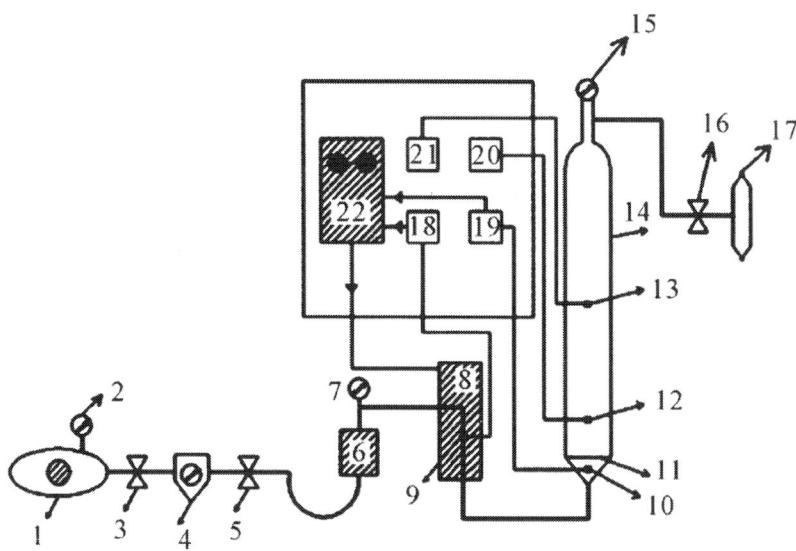

Figure 2: Experimental setup.

1. Compressor; 2. Gage pressure; 3. Flow control valve; 4. Pressure regulator; 5. Flow control valve; 6. Flow meter; 7. Gage pressure; 8. Electrical heater; 9. Heater temperature sensor; 10. Inlet gas temperature sensor; 11. Distributor plate; 12. Solid temperature sensor; 13. Outlet temperature sensor; 14. Glass column; 15. Gage pressure; 16. Flow control valve; 17. Orifice plate; 18. Heater temperature recorder; 19. Inlet gas temperature recorder; 20. Solid temperature recorder; 21. Outlet gas temperature recorder; 22. Temperature controller.

Table 1: Physical and geometrical properties of the bed and the seed

[6] column diameter	0.03 m
Bed height	0.2 m
Particle sphericity	0.91
Particle diameter	1.78 mm
Solid density	1145.77 kg/m³

Table 2: Operational conditions of experiments

Test no	u (m/s)	L (m)	T_{so}(°C)	T_{20}(°C)	M (kg)
1	1.57	0.040	23	47.7	0.012
2	1.57	0.040	22.5	42.5	0.012
3	1.57	0.040	24.1	38.5	0.012
4	1.96	0.050	24.1	56.5	0.012
5	2.36	0.057	24.1	57.5	0.012

Table 3: Descriptive Statistics of variation

variable	Mean	Median	Min	Std. Deviation	Mox
T^s	42.43	40.95	22.5	10.24	54.9
T_{gl}	45.67	45.7	33	7.23	55.4

50% of tests outlet gas temperature was lower than 45.7 and temperature variation coefficient for solid particles was 24% and variation coefficient for outlet gas was 16%.

Correlation Coefficients

According to Table 4 it is clarified surely that 99% of solid particle temperature has had meaningful relation with primary parameters and surely 99% of outlet gas has meaningful relation with primary parameters (P < 0.01). Also the relation between solid particles temperature and time is 99% meaningful surely (P < 0.05).

Table 5 is obtained by Stepwise method. After elimination of test beginning time, following results obtained: in the first step of Regression, inlet gas temperature variable which has the highest meaningful correlation with solid particle temperature has been entered in to model, in second step time variable, in third step gas velocity and in fourth step solid particles primary temperature have been added to Regression model. As it is obvious in Table 5 in forth step, F value has been calculated 1072.9 for Regression model and obtained which is meaningful in a = 0.01 level (P = 0.000 < 0.01), in other word solid particles temperature is predictable 99% by the use of primary parameters and time surely. Also in final step t certain number has been calculated for Regression coefficients in a = 0.01 which had been meaningful and meaninglessness assumption about Regression coefficients are rejected, so Regression model could be written as below:

Table 4: Pearson Correlation between solid particles and outlet gas temperature, initial parameters and time

	Pearson	T_s	T_{gl}	T_{s_0}	T_{g_0}	u_0	Time
	Correlation Coefficient	1	**0.84	**0.57	**0.57	**0.65	*0.37
T_s	P-Value. 2-tailed		0.000	0.000	0.000	0.000	0.020
	Correlation Coefficient	**0.84	1	**0.83	**0.84	**0.95	0.17
T_{gl}							
	P-Value. 2-tailed	0.000		0.000	0.000	0.000	0.296

T_{s_0}	Correlation Coefficient	**0.57	**0.83	1	**0.63	**0.86	0
	P-Value. 2-tailed	0.000	0.000		0.000	0.000	1.000
T_{g0}	Correlation Coefficient	**0.57	**0.84	**0.63	1	**0.87	0
	P-Value. 2-tailed	0.000	0.000	0.000		0.000	1.000
u_o	Correlation Coefficient	**0.65	**0.95	**0.86	**0.87	1	0
	P-Value. 2-tailed	0.000	0.000	0.000	0.000		1.000
Time	Correlation Coefficient	*0.37	0.17	0	0	0	1
	P-Value. 2-tailed	0.020	0.296	1.000	1.000	1.000	*0.37

**$P < 0.01$, *$P < 0.05$.

Table 5. Regression correlation between solid particles temperature with initial parameter and time

Step	Variable	Coefficients	Statistic t	P-Value	Statistic F	P-Value	R Square	Durbin-Watson accept/reject
1	constant	-6866.3	-1.9	0.062				
	$T_{s_0}^3$	0.871	**32.8	0.000	**1072.9	0.000	0.970	
2	constant	-149728.9	**_73	0.000				
	$T_{g_0}^3$	0.871	*15.5	0.000	**1350.1	0.000	0.988	
	$\sqrt{Tanh(time)}$	145901.4	**7.1	0.000				
3	constant	-149136.9	**_73	0.000				
	$T_{g_0}^3$	0.812	**259	0.000	*9009.1	0.000	0.990	1.55 accept
	$\sqrt{Tanh(time)}$	145901.4	**7A	0.000				
	u^3	1037.6	*2.2	0.036				

constant	−402397.2	** −5.7	0.000				
$T_{g_0}^3$	0.566	***7.9	0.000				
$\sqrt{\text{Tanh(time)}}$	145901.4	**8.9	0.000	**1072.9	0.000	0.993	4
u^3	2846.6	***4.5	0.000				
T_{S_0}	11713.1	***3.7	0.001				

**P < 0.01, *P < 0.05.

$$T_s^3 = -402397.2 + 0.566T_{g_0}^3$$
$$+ 145901.4\sqrt{\text{Tanh}(\text{time})} + 2846.6u^3$$
$$+ 11713.1T_{s_0}$$

(1)

Also 99/3% of solid particles temperature prediction are provided by the implementation of primary parameters and time to analyze residual independency, DurbinWatson statistic is used on a way that if this statistic rate is more than 1.5 and lower than 2.5 we can accept residuals independence. Because Durbin-Watson statistic is obtained as 1.55 we can conclude that residuals independency assumption is acceptable. To analyze residual normalization, Kolmogrov-Smirnov test has been used meaningful level was 0.274 about this test which shows accepted assumption about residuals normalization. In Figure 3 you can see results of simulation for solid particles temperature beside experimental data which are drawn for tests 1, 4.

Note Mentioned figure is made by stepwise method application. In first step of Regression, inlet gas primary temperature variable which has the highest meaningful correlation which solid articles temperature is entered in to the model, in second step time variable, in third step velocity variable and in fourth step temperature variable for solid particles are added to Regression model. As it is obvious in Table 6 F rate is calculated in fourth step for Regression model and 296/2 has been obtained as the result which is meaningful at a = 0/01 (P = 0.000 < 0.01), in other word with 99% assurance, outlet gas temperature is predictable by the use of primary parameters and time. Also in final step t calculated value is meaningful for Regression coefficients at a = 0.01 and Regression coefficient meaninglessness is rejected, so it is possible to write Regression model as below:

$$T_{gl}^3 = -430520.06 + 3549.8T_{g_0}$$
$$+ 40078.2 \, \text{Tanh}(\text{time}) + 35422.4u$$
$$+ 11332.1T_{s_0}$$

(2)

Also 97.1% of outlet gas prediction is determined by time and primary parameters. Because Durbin-Watson statistic is 1.53, we can conclude that residuals independency is accepted. Also meaningfulness level for Kolmogrov-Smirnov test is 0.086 which shows residuals normalization assumption has been accepted at a = 0.01 in Figure 4 you can see results about outlet gas simulation beside experimental data for test 1, 4.

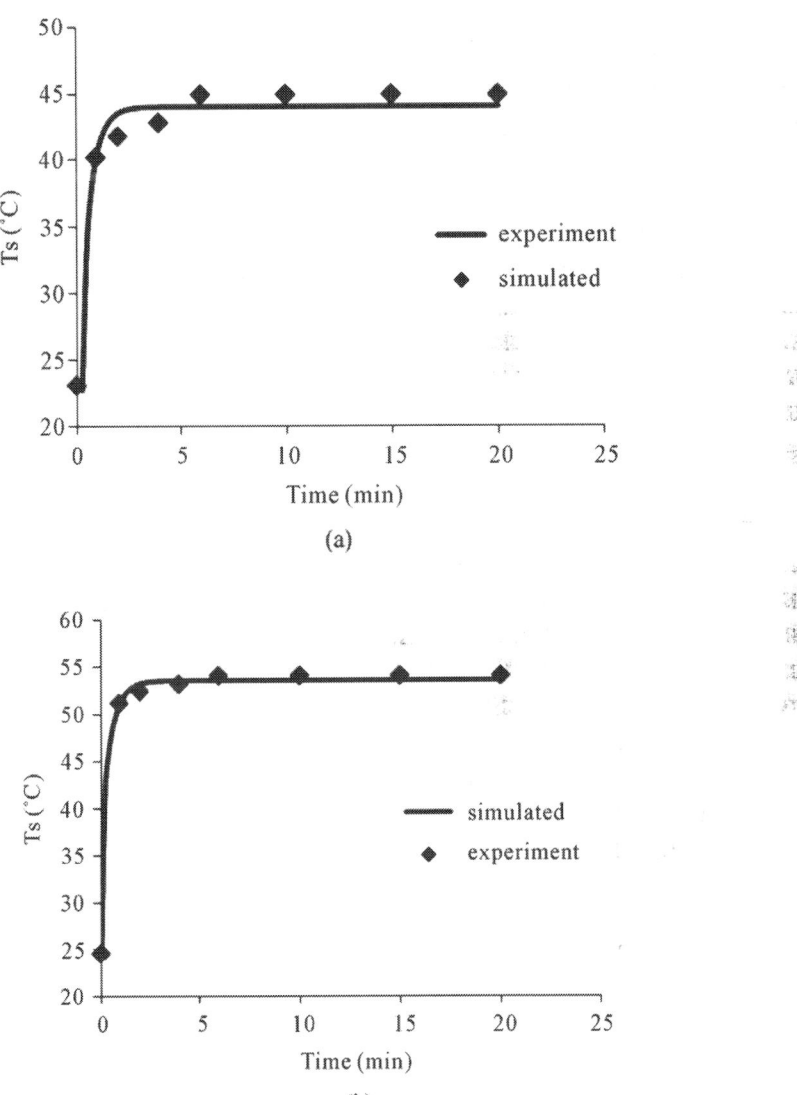

(a)

(b)

Figure 3: Comparison between experimental and simulated data for solid particle temperature: (a) Test #1; (b) Test #4.

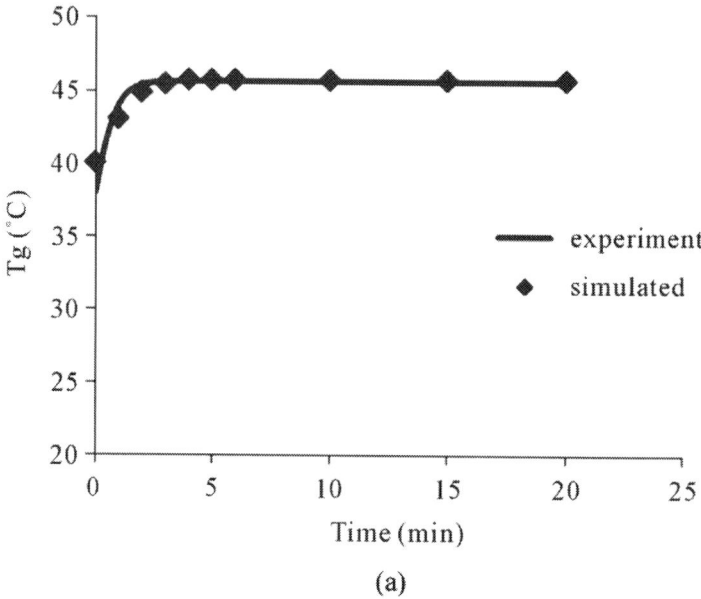

(a)

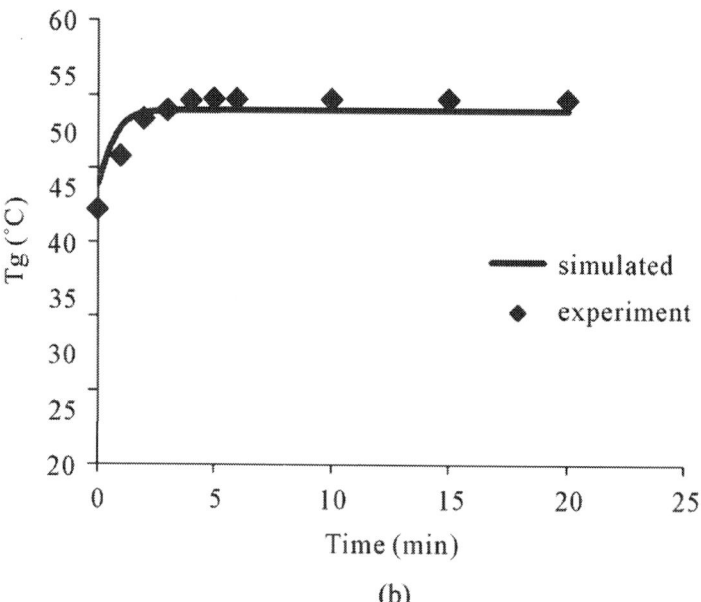

(b)

Figure 4: Comparison between experimental and simulated data for outlet gas temperature: (a) Test #1; (b) Test #4.

Table 6: Regression correlation between outlet gas temperature with initial parameter and time

Step	Variation	Coefficients	Statistic t	P-Value	Statistic F	P-Value	R Square	Durbin-Watson	Accept/Reject
1	Constant	-179416.2	**-10.5	0.000					
	T_{glo}	5803.5	"16.6	0.000	"276.7	0.000	0.879		
2	Constant	-213106.6	**-19.6	0.000					
	T_{glo}	5803.5	"28.1	0.000	"430.8	0.000	0.959		
	Tanh(time)	40078.2	***8.5	0.000					
3	Constant	-214596.4	**-21.1	0.000					
	T_{glo}	4974.6	"13.1	0.000	"331.1				
						0.000	0.965	1.53	Accept
	Tanh(time)	40078.2	"9.1	0.000					

u	23128.7	*2.5	0.016			
Constant	-430520.6	**_5.5	0.000			
T_{glo}	3459.8	**5.7	0.000			0.971
Tanh(time)	40078.2	**9.9	0.000	**296.2	0.000	
	35422.4	**3.7	0.001			
T_{s_0}	11332.1	**2.8	0.009			

CONCLUSIONS

In this research we have used an experimental setup to evaluate initial parameters effects on temperature variation process in solid-gas fluidized bed dryers in which a fluidization regime is made by air and solid combination and 5 series of test performed to analyze bed internal temperature variation and in every test inlet gas temperature is kept steady with high precision and solid particles and outlet gas temperature were registered during time, then Regression method applied to analyze experimental data. Results obtained from this study shows: 1) The maximum solid particles temperature variation and outlet gas temperature variation occur at the beginning of fluidization which show high heat transfer in these kinds of dryers. 2) Because of high temperature variations curve declivity, exponential functions are used to predict temperature variation in Regression model. Equations precision is very high to the point Regression equation given for solid particles temperature variation bas 99% conformity with experimental data precisely and presented equation accuracy for outlet gas temperature is about 97%. 3) Solid particles temperature variation velocity during time is more than variations related to outlet gas. 4) After passage from unsteady condition, solid particles temperature and outlet gas temperature from bed inclined to each other. 5) The most effective parameter on heat transfer, is the inlet gas temperature to bed and solid particles primary temperature; and inlet gas velocity effect is lower.

REFERENCES

1. K. Diazo and L. Octave, "Fluidization Engineering," 2nd Edition, Butterworth, Heinemann, 1991.

2. V. João, F. A. Biscaia Jr., C. Evaristo and M. Giulio, "Modeling of Biomass Drying in Fluidized Bed," Proceedings of the 14th International Drying Symposium, São Paulo, 22-25 August 2004, pp. 1104-1111.

3. H. Subramani, M. B. Mothivel and M. Lima, "Minimum Fluidization Velocity at Elevated Temperatures for Geldart's

Group-B Powders," Experimental Thermal and Fluid Science, Vol. 32, No. 1, 2007, pp. 166-173.

4.   A. C. Rizzi Jr., M. L. Passos and J. T. Freire, "Modeling and Simulating the Drying of Grass Seeds (Brachiaria brizantha) in Fluidized Beds: Evaluation of Heat Transfer Coefficient," Brazilian Journal of Chemical Engineering, Vol. 26, No. 3, 2009, pp. 545-554. doi: 10.1590/S0104-66322009000300010

5.   J. Khorshidi, H. Davari and F. Dehbozorgi, "Model Making for Heat Transfer in a Fluidized Bed Dryer," Journal of Basic & Applied Sciences, Vol. 1, No. 10, 2011, pp. 1732-1738.

6.   S. Minaei and E. Hazbavi, "Determination and Investigation of Some Physical Properties of Seven Variety Rapeseed," Iranian Journal of Food Science and Technology, Vol. 5, No. 4, 2008, pp. 21-28.

Chapter 6

Application of Photo-Fenton Process for the Treatment of Kraft Pulp Mill Effluent

M. D. Rabelo[1], C. R. Bellato[1], C. M. Silva[2], R. B. Ruy[1], C. A. B. da Silva[1], and W. G. Nunes[2]

[1]Department of Chemistry, Federal University of Viçosa, Viçosa, Brazil
[2]Department of Forest Engineering, Federal University of Viçosa, Viçosa, Brazil

ABSTRACT

The present work evaluated the use of photo-Fenton process for the treatment of kraft pulp mill effluent. The photo-Fenton best operating conditions, such as pH, concentration, and H_2O_2: Fe^{2+} ratio were evaluated. The efficiency of the treatment was measured by COD (chemical oxygen demand) removal. The results showed that the optimum pH for the photo-Fenton process was equal to 3. The increase in H_2O_2 application resulted in an efficiency increase of the photo-

Fenton process, although this was not a directly proportional relation. For most cases, the H_2O_2: Fe^{2+} proportion of 100:1 yielded the best results for COD removal. Solar radiation was more efficient than artificial UV to the COD removal. During the treatment the organic matter of the effluent was more oxidized than mineralized, showing a higher removal of COD than BOD (biochemical oxygen demand) and TOC (total organic carbon), respectively. So, photo-Fenton process increased the BOD/ COD ration but decreased the BOD/TOC ratio.

INTRODUCTION

Biological treatment processes are commonly used to treat pulp mill effluents and, in some cases, are not enough to meet regulations. In part, this is due to a fraction of recalcitrant organic matter present in the effluent, which is inert to biological oxidation.

An alternative to increase the removal of recalcitrant organic matter would be a pretreatment of this material, changing it into biodegradable compounds. The technology known as advanced oxidation processes (AOP) has been tested for the oxidation of some types of organic compounds, such as chlorinated phenols, being able to completely convert them into CO_2 or partially oxidize them [1] -[3] .

This type of treatment may also fractionate complex molecules of high molecular mass into simpler intermediate compounds, such as acetic, maleic, and oxalic acids, acetone and chloroform. These new formed compounds are part of the bioenergetic cycle of living organisms and, therefore, they are compatible with the biological treatment [4].

The AOP constitute a set of techniques based on the generation of the free radical ($\bullet$OH), in sufficient quantity to reach the oxidative degradation of the organic contaminants in water and wastewater [5]. The radical $\bullet$OH has a high potential for reduction, making it effective in oxidizing organic compounds.

Although several studies are being carried out, evaluating the oxidation of recalcitrant compounds in synthetic effluents, the application of the photo-Fenton process for the treatment of bleached kraft pulp mill effluent was not yet fully studied. Therefore, the main objective of this research was to evaluate the application of this process to the treatment of bleached kraft pulp mill effluent. Specific objectives were: i) to find the best operating pH; ii) to optimize the dosage and find the best H_2O_2: Fe^{+2} ratios; iii) to evaluate the source of radiation (solar and artificial); iv) to evaluate the efficiency of photo-Fenton reaction in removing organic contaminants from the industrial effluent.

MATERIAL AND METHODS

Effluents

The effluent used was collected at a bleached eucalyptus kraft pulp mill where D (EOP) DD bleaching sequence is used. The final effluent is a mixture of acid and alkaline filtrates from the bleaching plant and wastewater from other areas of the plant.

Photo-Fenton Treatment

The optimum pH to treat the effluent was established through preliminary testing to be in the range of 2.5 to 5.0. The treatments were carried out at two sources of radiation: solar and artificial. A 9-W black light lamp, with emission spectrum in the region of the near-to-visible ultraviolet (UVA) was used as source of artificial radiation. The experimental unit for the treatments carried out under sunlight consisted of four identical chemical photoreactors, with 0.043 m^2 of superficial area exposed to sunlight. A volume of 250 ml of effluent were treated in each photoreactor. The effluent layer thickness of 0.7 cm was used.

Fenton reaction was promoted by adding hydrated ferrous sulphate ($Fe_2SO_4 \cdot 7H_2O$) and stock solution of H_2O_2 to the effluent. The mixture was homogenized during the whole treatment by agitating with a magnetic bar. Each treatment lasted 120 minutes, all of them having been carried out between noon and 2 p.m. at ambient temperature.

The consumption of H_2O_2 during Fenton reaction was evaluated in parallel. For this purpose, a fifth treatment was required to be carried out concomitantly, by keeping the same conditions of the other four repetitions, but using a photoreactor with capacity for a higher volume of effluent (1250 ml). The concentration of H_2O_2 was determined by iodometry.

The treatments that used a black light lamp as source of radiation were carried out on a laboratory bench apparatus. The conditions and methods adopted for these treatments were analogous to those described for the treatments under solar radiation, except for the configuration of the photoreactor used. A configuration similar to those described by other authors [3] was used for these treatments.

A peristaltic pumps (Gilson, Milliplus 3), with flow adjusted so as to apply the whole dosage of reagent in the course of 60 minutes, was used for the treatments with continuous addition of reagents. The dosage of H_2O_2 and Fe^{2+} in each treatment corresponded to 500 and 50 mg·l^{-1}, respectively.

Chemical Analyses

The analyses of pH, COD, BOD, colour, total phenols, total nitrogen, total phosphorus, and chlorides, were conducted in accordance with the Standard Methods for the Examination of Water and Wastewater [6].

For the analysis of TOC, the samples were filtered in qualitative filter paper (quick filtration, black stripe), the pH having been adjusted thereafter to a range of 2 to 3 with 20% sulphuric acid. Later, TOC was analyzed in a TOC-5000A SHIMADZU apparatus according to the manufacturer's manual. The analyses of AOX (adsorbable organic halogens) of the effluents were done in Euroglas ECS 1600 equipment in accordance with SCAN standard [7].

RESULTS AND DISCUSSION

Effluent Characterization

The effluent used in the study showed typical characteristics of eucalyptus kraft pulp mill effluents (Table 1), as the high chloride content and the high ratio between organic matter (BOD or COD) and the nutrients nitrogen and phosphorus. The pH near the neutral and the BOD/COD ratio above 0.4 indicates biological treatability.

Selection of the Best pH for the Photo-Fenton Process

The mixture of all currents of sectorial effluents of a mill generates a final effluent, the pH of which normally ranges from 5 to 8, which is more suitable for the biological treatment, but incompatible with the photo-Fenton reaction (Figure 1).

The difference between the pH of the final effluent and the optimum pH found for the photo-Fenton process is an inconvenience for applying this technology. In operating conditions it would require large amounts of acid to lower the pH of the effluents to the optimum value for treatment.

Sources of Radiation for the Photo-Fenton Process

The continuous treatment of effluents by means of the photo-Fenton process requires the use of an artificial source of light for night periods. However, once the pH of the effluents is adjusted and the dosages of reagents are optimized, the levels of removal of COD reached in the conventional (biological) treatments can be attained and even surpassed by the photo-Fenton process, under solar or artificial radiation (Figure 2).

The concentrations of H_2O_2 at the beginning of the treatments of the sets of Experiments 1, 2, 3, 4, 5, and 6 were 60, 80, 100, 500, 1000, and 2000 mg·l^{-1}, respectively. The initial COD of the effluents ranged from 920 to 1190 mg·l^{-1}. The error bar shown in each column represents the standard deviation around an average of 4 repetitions. For each set of experiments, the averages followed by at least the same letter do not differ from each other at the level of a 5% probability by Tukey test. The average solar luminous intensity accumulated during the treatments was 475 ± 188 W·m^{-2}.

For the experiments carried out under solar radiation it became evident that the treatments employing 100:1 (H_2O_2:Fe^{2+}) ratio distinguished themselves from the remaining ones. Except for the results obtained for Experiment 1, the treatments that used 100:1 ratio presented the best performance in terms of COD removal. It is wellknown that radicals •OH generated in Fenton reaction react more rapidly with Fe^{2+} (rate constant = 3.0×10^8 $l \cdot mol^{-1} \cdot s^{-1}$) ions than with H_2O_2 (rate constant = 2.7×10^7 $l \cdot mol^{-1} \cdot s^{-1}$). Thus, it is probable that from the treatments carried out with the same concentration of hydrogen peroxide those having used lower iron loads, that is, 100:1, were less impaired by this type of deleterious reaction. On the other hand, the photoreduction of Fe^{3+} by sunlight maintains the concentration of Fe^{2+} in the reactive medium, guaranteeing the efficiency of the photo-Fenton process even in treatments with low initial concentration of ferrous ions [8].

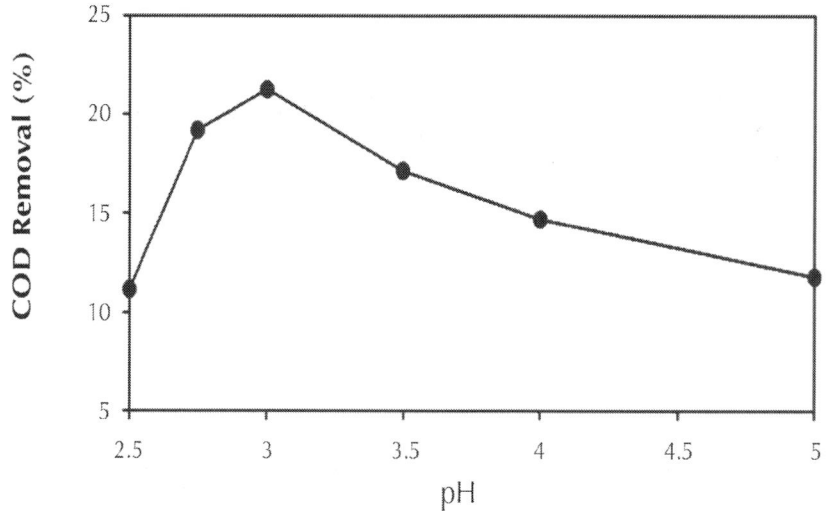

Figure 1: Effect of pH on removal of organic matter (COD) from the effluent through the photo-Fenton process. $[H_2O_2] = 100$ $mg \cdot l^{-1}$; $[Fe^{2+}] = 20$ $mg \cdot l^{-1}$; and initial COD = 1197 $mg \cdot l^{-1}$.

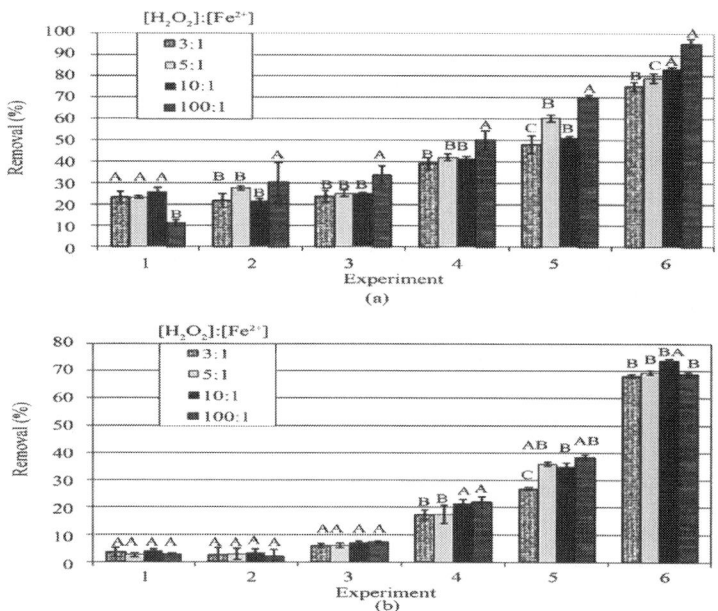

Figure 2: Removal of COD from the effluents treated by the photo-Fenton process, using solar (a) and ultraviolet light (b).

Table 1: Results of the physicochemical effluent characterization

Parameter	1st sampling* (mg·l−1)	2nd sampling* (mg·l−1)
Chloride	420 ± 26	486 ± 32
Colour	253 ± 19	238 ± 19
BOD5	694 ± 28	576 ± 75
COD	1235 ± 34	1204 ± 13
N total	0.76 ± 0.04	0.66 ± 0.06
P total	1.36 ± 0.01	1.37 ± 0.03
pH	7.4	7.2

*Average of 3 repetitions ± standard deviation.

These results disclose another particularly important aspect for a possible industrial application, since high iron loads, besides making the process more expensive, cause a thin sludge of iron hydroxide precipitates to form during neutralization of the effluent pH.

Previous works were carried out, comparing experiments at different initial concentrations of Fe^{2+}. Such works revealed that higher concentrations of Fe^{2+} at the beginning of Fenton reaction imply higher velocity of degradation of the organic contaminants, but in some cases without appreciably increasing the efficiency of the process [2] [8] -[10] . Higher concentrations of Fe^{2+} at the beginning of the reaction favour the formation of radicals •OH, which explains the higher velocity of degradation at the beginning of the reaction. On the other hand, higher concentrations of Fe^{2+} also increase the deleterious effect of this metal on the free radicals, contributing to reduce the efficiency of the photo-Fenton process.

Another important aspect was the high removal of COD reached in the Experiment 6, which attained 95.4% in the treatment that used the 100:1 ratio. In that treatment, the COD of the effluent was reduced to 43 $mg·l^{-1}$, a value far below the emission standards stipulated by rigorous environmental legislations. These observations corroborate the potential of AOP for the treatment of effluents.

In general, the treatments carried out with artificial irradiation were less efficient than those performed using solar irradiation, as far as the percentages of reduction in the COD of the effluents are concerned. As most experimental conditions were similar, the main explanation for such behaviour is that the black light lamp does not present the same efficiency as sunlight, as far as the photoreduction in Fe^{3+} is concerned. Furthermore, another factor that may have contributed to such results was the thickness of the effluent layer. A thicker layer may make it more difficult for the UV rays to penetrate the full extent of the sample, restricting photoreduction of Fe^{3+} to the most superficial layers of the liquid.

When statistically analyzing the results shown in the graphs of Figure 2, it is possible to conclude that the removal of COD increases by a lower proportion than the increase in the initial dosage of hydrogen peroxide. As the concentration of reagents (H_2O_2 and Fe^{2+}) is increased, the harmful effects are intensified, since H_2O_2 reacts with radicals • OH, competing with the organic matter oxidation [11] [12].

Figure 3 shows the decrease in concentration of H_2O_2 during effluent treatment. The initial concentration of H_2O_2 in the treatments was equal to: a) 60 $mg·l^{-1}$, b) 80 $mg·l^{-1}$, c) 100 $mg·l^{-1}$, d) 500 $mg·l^{-1}$, e) 1000 $mg·l^{-1}$, and f) 2000 $mg·l^{-1}$. The ratios 3:1, 5:1, 10:1, and 100:1

between reagents H_2O_2 and Fe^{2+}, adopted in each treatment, are shown in the area of each graph.

The decrease in concentration of H_2O_2 as a function of the treatment time reveals that the most pronounced consumption of this reagent occurs during the first minutes of reaction, except for the treatments having used the 100:1 proportion. In these treatments, the consumption of H_2O_2 is well-distributed during the whole reaction time, probably because the lower availability of ferrous ions (required to form the radicals • OH in the reactive medium reduces the velocity of decomposition of the hydrogen peroxide. In the treatments using Fenton reagent in the proportions of 3:1, 5:1, and 10:1, for which a quick consumption of H_2O_2 was observed, it is probable that the generation of radicals • OH and consequently the organic matter degradation mainly occur at the beginning of the treatment.

From the operating point of view, this is a further important aspect, as reduction in effluent treatment time can be achieved by increasing the dosage of Fe^{2+} at the beginning of the treatment. Thus, for the design of an effluent treatment plant by means of the photo-Fenton process the hydraulic retention time should be determined as a function of the ratio between reagents H_2O_2 and Fe^{2+} to be used.

Effluent Treatment Efficiency

A more complete evaluation of the performance of the photo-Fenton process was carried out as soon as the most suitable ratio (100:1) between the dosages of reagents H_2O_2 and Fe^{2+} was defined.

The results showed in Table 2 indicate that the removal of COD by the photo-Fenton process increased as the dosages of reagents H_2O_2 and Fe^{2+} increased. Nevertheless, the increase in COD removal did not reflect the same proportion of the increase in the initial dosage of the reagents.

The removal of COD surpassed removals of BOD, TOC, and AOX. Apparently inconsistent, this result can be understood, since during the photo-Fenton reaction a good part of the organic matter is not completely mineralized (converted into CO_2). A part of it turns into partially oxidized compounds, which reduces COD, and/or is converted into simpler subproducts, such as the carboxylic and formic acids and aldehydes, which are, on the other hand, more biodegradable

[2] [13] . The results also show that BOD/TOC ratio, understood as the biodegradability per carbon atom, decreased after effluent treatment, which reinforces the idea that the organic matter was modified (partially oxidized), rather than mineralized.

The BOD/COD ratio of the effluents increased after two out of the three treatments, characterizing an indication of increase in biological treatability of the effluent. This aspect of the photo-Fenton process suggests that there is the potential to improve the performance of a subsequent biological treatment.

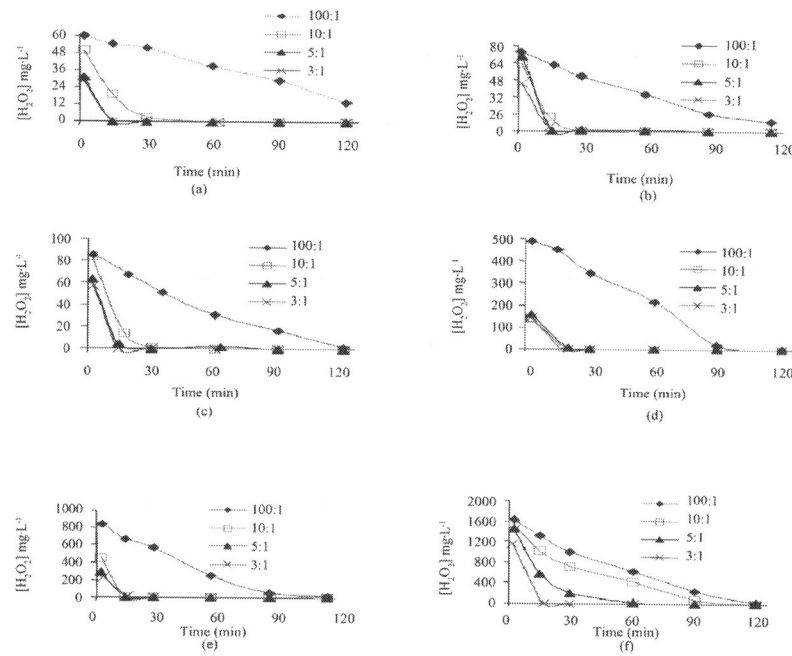

Figure 3: Decrease of the concentration of H_2O_2 during effluent treatment. The initial concentrations of H_2O_2 were: (a) 60 mg·l^{-1}; (b) 80 mg·l^{-1}; (c) 100 mg·l^{-1}; (d) 500 mg·l^{-1}; (e) 1000 mg·l^{-1}; and (f) 2000 mg·l^{-1}.

Table 2: Removal of COD, BOD, TOC, and AOX from effluents treated by using 100:1 ratio between reagents H_2O_2 and Fe^{2+}

	COD mg·l−1	BOD5 mg·l−1	TOC mg·l−1	AOX mg·l−1	BOD5/ COD	BOD/ TOC
Raw Effluent	1246	502	457	10.7	0.40	1.10
Treated Effluent [H2O2] = 80 mg·l−1	988	441	407	9.8	0.45	1.08
Removal (%)	20.7	18.1	11.0	8.3	-	-
Treated Effluent [H2O2] = 500 mg·l−1	816	323	384	9.1	0.40	0.84
Removal (%)	34.5	35.7	16.0	14.4	-	-
Treated Effluent [H2O2] = 1000 mg·l−1	476	272	345	8.6	0.57	0.79
Removal (%)	61.8	45.8	24.4	19.1	-	-

The reductions in AOX and TOC are in the same range, which means to say that the intermediate compounds formed are still chlorinated.

CONCLUSIONS

The high removals of COD, obtained in some experiments, show that the photo-Fenton process has the technical potential to be used in the treatment of pulp mill effluents.

The pH 3 was the optimum for application of the photo-Fenton process to the treatment of effluents from the bleached kraft pulp industry. At this pH, the experiments carried out with initial ratio of 100:1 between the concentrations of reagents H_2O_2 and Fe^{2+} presented the highest removals of COD. The removal of COD by the photo-Fenton process increases as the dosages of reagents H_2O_2 and Fe^{2+} are increased. Nevertheless, the increase in COD removal does not reflect the same proportion of the increase in the initial dosage of the reagents. The treatments carried out under sunlight presented higher reductions in COD than those performed with artificial radiation. Treatments carried out under sunlight, with application of 2000 and 20 $mg·l^{-1}$ of H_2O_2 and Fe^{2+}, respectively, removed over 95% of COD from the effluents. For the treatments performed with artificial irradiation, the highest reduction (73.5%) in COD was noticed when 2000 and 200 $mg·l^{-1}$ of H_2O_2 and Fe^{2+}, respectively, were used.

Experiments carried out under sunlight showed that removal of BOD, TOC, and AOX also increases as the dosages of Fenton reagent are increased, but in the same way as with regard to COD, it does not reflect the same proportion.

During the chemical treatment, the organic matter present in the effluents was oxidized, rather than mineralized, higher percentage removals of COD, BOD, and TOC, respectively, having been observed. For this reason, the photo-Fenton process increased BOD/COD ratio, but reduced BOD/TOC ratio.

ACKNOWLEDGMENTS

The authors would like to thank Capes, FAPEMIG, CNPq and Cenibra.

REFERENCES

1.   Titus, M.P., Molina, V.G., Banos, M.A., Gimenez, J. and Santiago, E. (2004) Degradation of Chlorophenols by Means of Advanced Oxidation Processes: A General Review. Applied Catalysis B: Environmental, 47, 219-256. http://dx.doi.org/10.1016/j.apcatb.2003.09.010

2.   Katsumata, H., Kawabe, X., Kaneco, S., Suzuki, T. and Ohta, K. (2004) Degradation of BispHenol A in Water by the Photo-Fenton Reaction. Journal of Photochemistry and Photobiology A, 162, 297-305. http://dx.doi.org/10.1016/S1010-6030(03)00374-5

3.   Ghaly, M.Y., Hartel, G., Mayer, R. and Roland, H. (2001) Photochemical Oxidation of P-Chlorophenol by UV/H2O2 and Photo-Fenton Process. A Comparative Study. Waste Management, 21, 41-47. http://dx.doi.org/10.1016/S0956-053X(00)00070-2

4.   Bigda, R.J. (1995) Consider Fenton's Chemistry for Wastewater Treatment. Chemical Engineering Progress, 91, 62-66.

5.   Legrini, O., Oliveiros, E. and Braun, M. (1993) Photochemical Processes for Water Treatment. Chemical Reviews, 93, 671-698. http://dx.doi.org/10.1021/cr00018a003

6.   (1998) Standard Methods for the Examination of Water and Wastewater. 20th Edition, American Public Health Association/ American Water Works Association/Water Environment Federation, Washington DC.

7.   Scan: Scandinavian Pulp, Paper and Board (1994) Scantest Standard. Distribuition: Secretariat, Scandinavina Pulp, Paper and Board Test Committee, Stockholm, August 1994.

8.   Kavitha, V. and Palnivelu, K. (2003) The Role of Ferrous ion in Fenton and Photo-Fenton Processes for the Degradation of Phenol. Chemosphere, 55, 1235-1243.http://dx.doi.org/10.1016/j. chemosphere.2003.12.022

9.   Pérez, M., Torrades, F., Hortal, J.A.G., Domenech, X. and Peral, J. (2002) Removal of Organic Contaminants in Paper Pulp Treatment Effluents under Fenton and Photo-Fenton Conditions. Applied Catalysis B: Environmental, 36, 63-74.http://dx.doi.org/10.1016/ S0926-3373(01)00281-8

10. Moraes, J.E.F., Silva, D.N., Quina, F.H., Filho, O.C. and Nascimento, C.A.O. (2004) Utilization of Solar Energy in the Photodegradation of Gasoline in Water and of Oil-Field-Produced Water. Environmental Science Technology, 38, 3746-3751.http:// dx.doi.org/10.1021/es034701i

11. Rodriguez, M., Sarria, V., Esplugas, S. and Pulgarin, C. (2002) Photo-Fenton Treatment of a Biorecalcitrant Wastewater Generated in Textile Activities. Journal of Photochemistry and Photobiology A: Chemistry, 151, 129-135. http://dx.doi. org/10.1016/S1010-6030(02)00148-X

12. Torrades, F., Pérez, M., Mansilla, H.D. and Peral, J. (2003) Experimental Design of Fenton and Photo-Fenton Reactions for

the Treatment of Cellulose Bleaching Effluents. Chemosphere, 53, 1211-1220. http://dx.doi.org/10.1016/S0045-6535(03)00579-4

13.   Neyens, E. and Baeyens, J. (2003) A Review of Classic Fenton's Peroxidation as an Advanced Oxidation Technique. Journal of Hazardous Materials, B98, 33-50.http://dx.doi.org/10.1016/S0304-3894(02)00282-0

Tribological Performance of Ptfe- and Peek-based Coatings Under Oil-less Compressor Conditions

Seung Min Yeo[a] and Andreas A. Polycarpou[a, b]

[a]Department of Mechanical Science and Engineering, University of Illinois at Urbana-Champaign, 1206 W. Green Street, Urbana, IL 61801, USA

[b]Department of Mechanical Engineering, Khalifa University of Science, Technology and Research, Abu Dhabi, UAE

ABSTRACT

Due to favorable tribological performance, PTFE- and PEEK-based polymeric coatings have received interest in air-conditioning and refrigeration compressor applications, as a potential solution to supplement and potentially replace conventional oil lubricants. The literature in this area is somewhat scarce, especially on the tribological performance of PTFE- and PEEK-based polymeric coatings under aggressive conditions simulating compressor operation. In this work, several PTFE-, PEEK-, resin- and fluorocarbon-based polymeric coatings, coated on gray cast iron were tribologically evaluated using a specialized tribometer under compressor specific conditions, that included oscillatory motion (simulating piston-type compressors) and unidirectional motion (simulating swash plate compressor operation). The coatings showed good to excellent tribological performance, and in general PTFE-based coatings exhibited better friction and wear behavior than the rest of the coatings, including PEEK-based coatings. Long-term durability experiments also showed the superiority and suitability of PTFE/Pyrrolidone coating for potential use in oil-less compressors (where oil-less conditions refer to operation in the absence of any liquid lubricant).

INTRODUCTION

To meet higher performance requirements, modern air-conditioning and refrigeration compressors need to function at harsher operating conditions, including higher speeds and loads. Such severe conditions could cause higher friction, excessive wear and catastrophic failures of critical interacting components, rendering the device inoperable. A further complexity in the operation of such devices is that the state of liquid type lubrication is usually unknown, and is considered (at best) to be in the boundary/mixed (or starved) lubrication regimes [1]. Moreover, due to thermodynamically negative effects of lubricants on the refrigeration cycle of typical refrigeration and air conditioning compressors [2], recent research interest has focused on oil-less type compressors. Lastly, environmental concerns towards eco-friendly products, led the transition of refrigerants used in compressors from chlorofluorocarbon (CFC) to hydrofluorocarbon (HFC), to hydrocarbon (HC), and eventually to minimal or "zero" global warming refrigerants, such as CO_2[3]. With reduced lubrication in modern compressors and elimination of chlorine that was present in the CFC refrigerants (which formed protective ferrous chloride layers on iron surfaces [4]), it is necessary to explore protective surface coatings for critical tribopairs for reliable operation. Fig. 1 summarizes the motivation towards oil-less air-conditioning compressor operation via the use of some sort of protective tribological coatings.

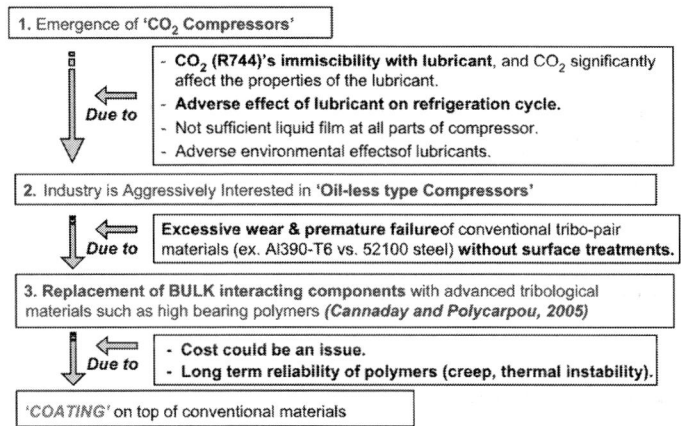

Figure 1: Motivation and development of advanced coatings for oil-less compressor applications.

Coatings can be broadly classified as either "hard" or "soft" coatings, with one category of soft coatings being polymer-based coatings. Conventionally, hard coatings such as diamond-like carbon (DLC), Ti–N and WC/C synthesized through physical vapor deposition (PVD) techniques are thought to be effective in preventing both abrasive and adhesive wear of metal sliding contacts [5]. DLC is one of the most researched tribological coatings, and is found in commercial applications such as magnetic storage hard disk drives [6] and in automotive applications. These coatings are in the form of a hard film on the surface that are able to reduce scratching, and offer good load-carrying capacity. Tribological coatings also have the ability to form low shear strength reaction layers and transfer layers on the top surface and the counterface, resulting in weak shear planes and thus low friction [7]. Another type of hard coatings, WC/C-based coatings were also shown to have superior tribological properties not only as far as wear resistance, but also low friction coefficient values as low as 0.05 under dry unidirectional pin-on-disk sliding conditions[8]. Hard coatings are relatively expensive and exhibit difficulties in coating them on substrates with low surface energy or high roughness [9]. Moreover, hard coatings sometimes could wear out the counterface they slide against, due to their relatively high hardness [10], and alternative solutions need to be explored.

Recent attention has focused on soft, thermoplastic-based polymer materials such as polytetrafluoroethylene (PTFE) and polyetheretherketone (PEEK). The bulk form of these materials shows relatively low friction coefficient and self-lubricating properties [11]. PTFE has been used extensively since its discovery because of its desirable tribological properties such as chemical inertness and superb lubricity[12]. However, bulk PTFE suffers from poor resistance to wear and creep, because it easily yields in shear due to its relatively low intermolecular strength [13]. Thus, PTFE is typically used in the form of composites, either (1) as a matrix filled with various hard fillers and micro/nano particles such as glass fibers, ceramics, MoS_2 and carbon nanotubes (CNTs) to enhance its wear resistance [14], or (2) as a filler into polymeric materials which have good wear resistance but poor frictional properties, such as PEEK, in which case it lowers its friction while retaining high wear resistance [15] and [16]. In fact, PEEK composites have been investigated as bearing and sliding materials for use in industrial applications due to their favorable tribological characteristics [17] and [18]. PEEK is a semi-crystalline high performance engineering polymer with good thermal (T_g=143 °C, T_m=338 °C, continuous service temperature=250 °C, and heat distortion temperature often in excess of 300 °C), as well as good mechanical properties (strength, modulus, toughness, and resistance to creep, abrasion, and fatigue) [17] and [19].

Compared to the abundance of research on the tribology of bulk polymers, there is little work on the tribological behavior of polymeric-based coatings (with thicknesses in the 10's of micrometers). The tribological behavior of polymeric coatings may not necessarily follow the same behavior as that of their bulk counterparts. This is partly due to the fact that the structure of the polymer materials changes during the coating process, which is known to affect the tribological performance, e.g., being amorphous or crystalline [20]. Research examining the tribological behavior of polymeric coatings [20], [21] and [22], under mild conditions of 1 N–10 N normal load and 0.25 m/s–2.5 m/s sliding speed have been reported. Tribological testing of commercially available PTFE- and PEEK-based polymeric coatings under realistic compressor operating conditions (4.5 m/s sliding speed and normal loads of 400 N–2000 N) has also been reported [23], [24] and [25]. Demas and Polycarpou [23] evaluated the tribological performance of two different PTFE/Pyrrolidone coatings and a Resin/PTFE/MoS_2

coating under unlubricated and carbon dioxide (CO_2) refrigerant environments with reciprocating motion, simulating the wrist pin in a piston-type compressor. The PTFE-based coatings showed excellent frictional characteristics, with friction coefficient values as low as 0.1, and good wear resistance attributed to the beneficial effects of the generated wear debris at the interface. Nunez et al. [25] performed a comparative testing of PEEK-based coatings under starved lubricated (mixture of R-134A refrigerant and polyalkylene glycol lubricant) and unidirectional sliding conditions simulating swash plate compressors. A study of a newly developed Aromatic Thermosetting Copolyester applied as a coating, has showed exceptionally low wear performance and reasonable friction behavior, and is under further development [26].

A research question that still remains unanswered is which of the two families of coatings; PTFE-based or PEEK-based offer superior tribological performance under either piston-type (reciprocating) or swash plate (unidirectional) compressor conditions, as well as their comparison with their bulk counterparts. In this work seven different PTFE- and PEEK-based coatings are tribologically evaluated under different operating conditions, simulating both unidirectional and oscillatory motions, and compared to bulk polymers.

EXPERIMENTAL

Oscillatory and unidirectional testing conditions (simulating piston-type and swash plate compressors) were performed. Photographs of typical piston-type and swash plate compressors and their corresponding critical tribocontacts are shown in Fig. 2. To simulate the contact geometry in a wrist pin/connecting rod interface of piston-type compressors, shown in Fig. 2(a), the cylindrical pins were cut to length and oriented to create a line contact with a reciprocating disk coated with different polymeric coatings. The photograph and illustration in Fig. 2(b) shows the contact configuration for unidirectional tests, showing a crown shaped pin (which is an actual component used in swash plate compressors and referred to as shoe in this work) in contact with a rotating disk coated with different polymeric coatings, simulating a swash plate compressor.

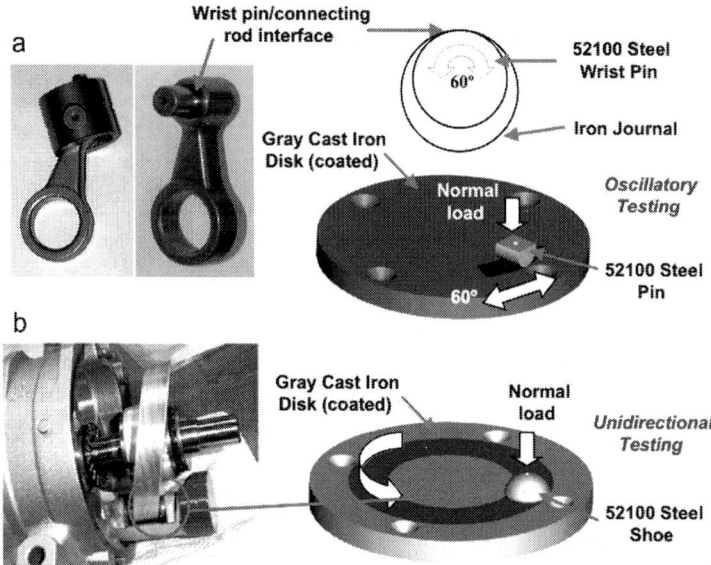

Figure 2: (a) Cylindrical self-aligned pin-on-disk test configuration for oscillatory testing simulating piston-type compressors and (b) crowned self-aligned pin (or shoe)-on-disk test configuration for unidirectional testing simulating swash plate compressors.

A specialized high pressure tribometer (HPT) was used to perform both types of experiments. The disk sample is fixed on the upper rotating spindle capable of unidirectional sliding speeds up to 2200 revolutions per minute (rpm) and reciprocating motion up to 4.85 Hz with variable oscillation amplitude. The temperature of the upper spindle and disk assembly can be regulated from –20 °C to 120 °C. The pin is placed on the lower fixture whose vertical position is adjusted by a mechanical power screw mechanism, enabling a controlled normal load ranging from 45 N to 4450 N applied on the pin-on-disk interface. Also, this lower fixture is mounted on a 6-axis force transducer so that the forces in the x, y, and z linear directions can be measured *in-situ* to obtain the coefficient of friction (COF). Both the upper spindle and lower fixture holding the samples are located inside an environmentally controlled vacuum chamber of the HPT capable of pressure control from near near-zero up to 1.72 MPa (250 psi).

Samples

Gray cast iron (Dura-Bar® G2), a commonly used material in compressors with a bulk hardness of 2.2 GPa, was machined to 75 mm diameter and 6.8 mm thickness disks as shown in Fig. 2. The disks had an initial (as machined) root-mean-square surface roughness (R_q) of 0.4 μm. After grit-blasting with aluminum oxide (which was performed before depositing the polymeric coatings to increase adhesion), R_q increased to 3.5 μm, which facilitated the deposition of polymeric coatings on the substrate surface. Seven different polymeric coatings, namely, PTFE/Pyrrolidone-1 (DuPont™ Teflon® 958-303), PTFE/Pyrrolidone-2 (DuPont™ Teflon® 958-414), Resin/PTFE/MoS$_2$ (Whitford Xylan® 1052), PEEK/PTFE (1704 PEEK/PTFE®), PEEK/Ceramic (1707 PEEK/Ceramic®), Fluorocarbon (Impreglon® 218), and PTFE/MoS$_2$(Fluorolon® 325) were deposited on the grit-blasted substrates using a spray gun. The name of each coating shows their base materials. The entire composition (including solvents) of these commercially available coatings is proprietary and thus unknown. However, some information is knows, e.g., in the case of DuPont™ coatings, they were dried for 5 min after application, and then baked for 15 min at 343 °C (650 °F). These coatings can be cured at temperatures as low as 177 °C (350 °F) by extending the cure time, but the toughness and durability of the coating decreases as the cure temperature is reduced below 343 °C. The entire deposition processes were performed by two authorized applicators, Orion Industries (for Teflon® and Xylan® coatings) and Southwest Impreglon (for PEEK, Impreglon® and Fluorolon®coatings). Further information on the application method for PTFE- and PEEK-based coatings can be found in Refs. [23] and [25] and from the aforementioned coating companies.

The surface roughness of each coating was measured using a stylus profilometer (Tencor P-15™) and are summarized in Table 1 along with the coating's hardness values measured using a micro-Vickers tester. The average surface roughness values of the coatings were in the range from 1.2 μm to 3.3 μm. As expected, the hardness of the polymeric coatings was found to be lower than the substrate hardness. The thickness of each coating was measured using cross section scanning electron microscopy (SEM); atypical such measurement is shown in Fig. 3. For all polymeric coatings, their thickness values, measured over 200 μm length were not uniform, and in the range of 20 μm–40 μm, which is much thicker than typical hard coatings such as diamond-like-carbon (DLC) and WC/C with a thickness of 2.5 µm [27].

Table 1: Mechanical properties of polymeric coatings, substrate, and pin used in this work

	Samples	Hardness (GPa)	Roughness, Rq (µm)
Coating	PTFE/Pyrrolidone-1	0.38	3.3
	PTFE/Pyrrolidone-2	0.25	1.2
	Resin/PTFE/MoS$_2$	0.32	2.3
	PEEK/PTFE	0.37	2.6
	PEEK/Ceramic	0.24	3.1
	Fluorocarbon	0.25	2.4
	PTFE/MoS$_2$	0.24	2.2
Substrate	Dura-Bar® G2, Gray case iron	2.2	0.4
Pin	52100 Steel	11.7	0.035

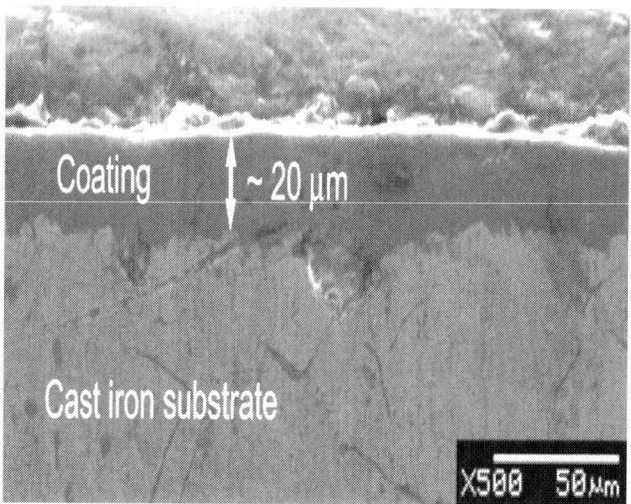

Figure 3: Representative cross-section SEM image showing the thickness of PTFE/Pyrrolidone-2 coating, on cast iron substrate.

The semi-cylindrical pins for the oscillatory experiments were made out of 52100 steel wrist pins, and were 8 mm in diameter and 8 mm long with a 1 mm diameter hole drilled up to 2 mm below the contact surface to accept a miniature thermocouple and measure the *in-situ* near contact temperature (NCT) during testing (Fig. 2(a)). The crowned 52100 steel shoes for the unidirectional experiments were 9.6 mm in diameter, and also had a 1 mm diameter hole for thermocouple insertion.

Tribological Testing Conditions

Before testing, the non-coated pin samples were ultrasonically cleaned in acetone for 10 min, rinsed with 2-propanol, and dried with a warm air blower. The polymeric coated disk samples were cleaned in 2-propanol and dried, but were not exposed to acetone.

For the oscillatory experiments simulating piston-type compressor operation (Fig. 2(a)), a constant normal load of 445 N was applied on the stationary pin (corresponding to a nominal mean contact pressure of 450 MPa), representing aggressive compressor conditions. To enable run-in and avoid initial abrupt failures, the pin was gently brought into contact with the disk surface with a 44.5 N preload before test initiation. The coated disk was reciprocating at a frequency of 4.5 Hz with 60° amplitude (peak-to-peak) and an average wear track diameter of 47.6 mm producing an average linear velocity of 0.22 m/s. To examine the applicability of these coatings for use in oil-less compressors, tests were performed at 172 kPa (25 psi) R744 (CO_2) refrigerant environment with no lubricant. Also, all tests were performed at room temperature (21 °C–23 °C) conditions without any temperature feedback control on the HPT, thus allowing natural increase of the temperature on both pin and disk samples (with continuous sliding during testing). Two tests were performed for each condition for repeatability, and each test was run for 30 min, corresponding to a sliding distance of 396 m. Tests were stopped earlier when the friction coefficient and near contact temperature abruptly increased, indicating destruction of the coating and sudden catastrophic failure of the interface. Under oscillatory testing conditions, four of the coatings (PEEK/PTFE, PEEK/Ceramic, Fluorocarbon, PTFE/MoS_2) were tested in this work, as the other three coatings were tested under identical conditions in earlier work [23] and were directly used in this work. For unidirectional testing, all seven coatings were tested.

For the swash plate compressor simulation, the unidirectional sliding tests were performed with a rotating speed of 1500 rpm and an average wear track diameter of 43.2 mm which corresponded to a linear speed of 3.75 m/s. Environmental conditions were the same as oscillatory testing (i.e., 445 N normal load, room temperature without any feedback control, 172 kPa of CO_2 refrigerant with no lubricant, and 30 min duration, corresponding to 6.75 km sliding distance). As the shoe is crowned the exact contact pressure is difficult to calculate and thus the contact load is used instead of the contact pressure. Similarly, two tests were performed for each condition, showing repeatable behavior. For the group of coatings which exhibited better tribological performance, additional unidirectional 3-h long duration (durability) tests were performed to examine their long-term behavior, simulating compressor life tests (the 3 h duration tests corresponded to 40.5 km sliding distance). Table 2 summarizes the experimental conditions for both oscillatory and unidirectional testing conditions.

Table 2: Summary of experimental conditions

	Oscillatory	**Unidirectional**
Temperature (°C)	21–23	
Environment (refrigerant)	25 psi of R744 (CO_2)	
Normal load (N)	445	
Contact geometry	Line contact	Crowned contact
Contact pressure (MPa)	450	–
Sliding speed (m/s)	0.22	3.75
Test duration	30 min	30 min, 3 h

Using a stylus profilometer (Tencor P-15™), four different line scans were taken across the wear tracks generated on the disk surfaces after testing, and an average value was recorded. From the line scan data, the exact wear volume loss of each coating could be precisely determined as described in Fig. 4. Then, the normalized wear rate for each coating was calculated by dividing the wear volume by the normal load and the total sliding distance.

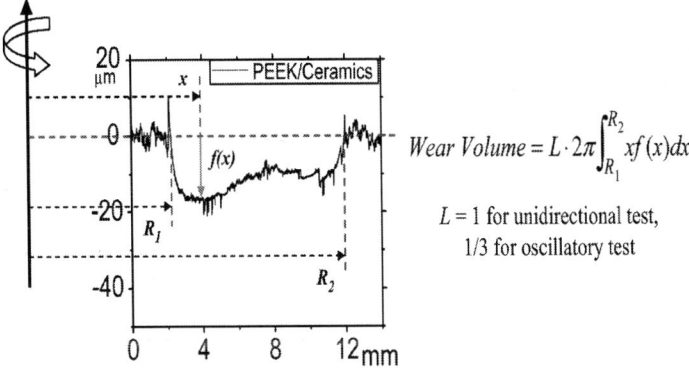

Figure 4: Typical wear track showing the calculation of the wear volume.

RESULTS AND DISCUSSION

Oscillatory Testing (Piston-type Compressors)

Representative friction coefficient and near contact temperature measurements of one of the polymeric coatings (PTFE/MoS$_2$) under dry-oscillatory conditions is shown in Fig. 5. Two different tests are shown, exhibiting similar behavior, thus validating repeatability. The friction coefficient gradually increased for the first 10 min, and then slightly decreased, reaching a steady-state value at 15 min. This friction behavior was explained in [23] by the third-body self-lubricating behavior of the generated wear debris particles. Initially during the running in period and as the coating is worn out, the friction coefficient increases because higher frictional force is needed to generate more wear particles. At some point, the amount of wear debris generated is sufficient to act as a solid lubricant, thus preventing further wear of the coating, which results in steady-state behavior. This friction coefficient behavior was similar for all coatings, even though their absolute values were different (as it will be discussed later). In regards to the near contact temperature behavior, it quickly increased from room temperature (at the start of the test) to about 45 °C during the first 1 min, and then followed the same trend as the friction coefficient.

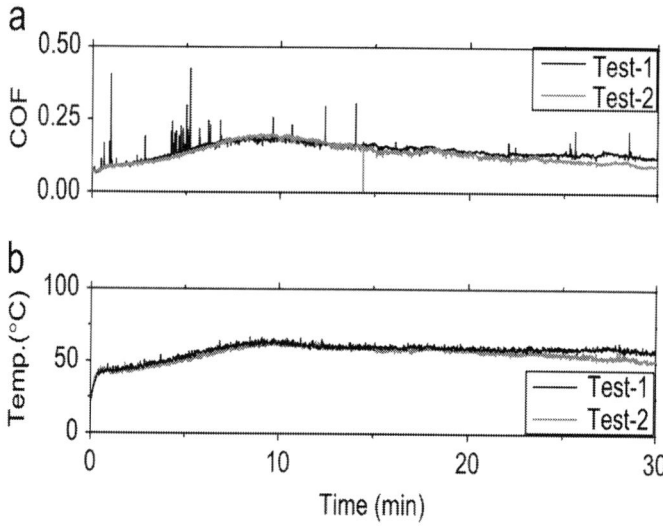

Figure 5: *In-situ* (a) friction coefficient and (b) near contact temperature of PTFE/MoS$_2$ coating during 30 min oscillatory testing. The sharp transients, especially in the friction data are electrical noise as the data is unfiltered.

Typical profilometric wear track measurements of the 4 coatings tested under dry-oscillatory conditions are shown in Fig. 6. The wear track profiles of the other three coatings under the same testing conditions can be found in Ref. [23]. Due to relatively high contact pressure (450 MPa) in this condition, wear of the coatings was observed for all coatings. Among the 4 coatings shown in Fig. 6, PTFE/MoS$_2$ showed the highest wear resistance, with an average wear depth of 17 μm. PEEK-based coatings exhibited higher wear, around 40 μm, and in the case of PEEK/Ceramic coating, over 45 μm of wear, which corresponded to the total coating thickness. Nevertheless, none of these coatings exhibited catastrophic failure during the test duration, due to the lubricity effect of the wear debris trapped inside the wear tracks. The wear rate for each coating was calculated using the method discussed in Fig. 4, and plotted on the *y*-axis along with the friction coefficient values on the *x*-axis, shown in Fig. 7.

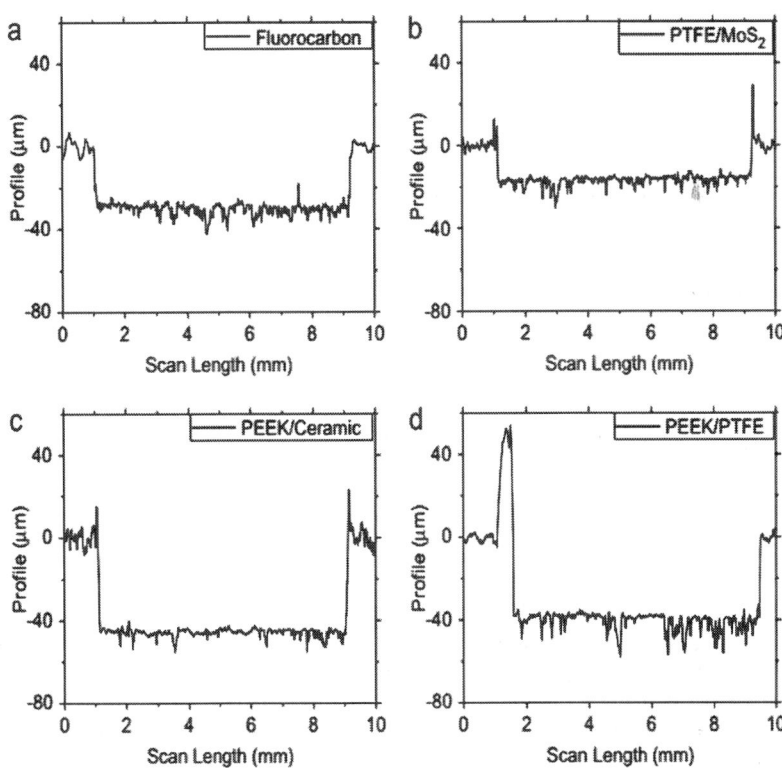

Figure 6: Typical profilometric wear track measurements of (a) fluorocarbon, (b) PTFE/MoS$_2$, (c) PEEK/Ceramic, and (d) PEEK/PTFE coatings after 30 min oscillatory testing.

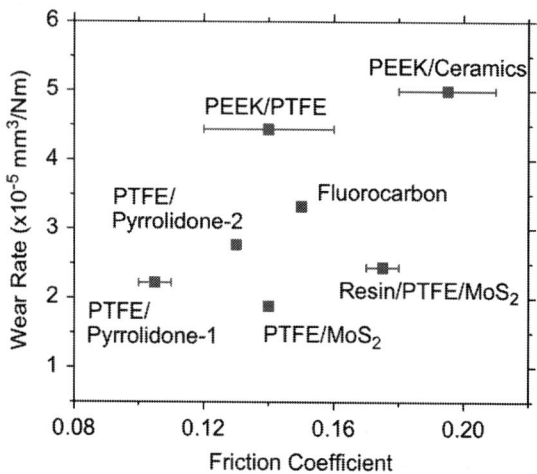

Figure 7: Friction coefficient (x-axis) vs. wear rate (y-axis) of seven polymeric coatings in CO_2 (22 °C and 25 psi) environment at 445 N normal load under dry-oscillatory conditions.

Referring to Fig. 7, all coatings showed relatively low friction coefficient values in the range of 0.1–0.2 while their wear rate was of the order of 10^{-5} mm³/N m, which is higher than that of hard coatings such as DLC and CrN coatings (which is in the range of 10^{-8} mm³/N m–10^{-9} mm³/N m [7]). Note however that the wear rate values for the polymeric coatings represent upper bound estimates since the experiments can run much longer times without further wear, as shown later. The friction and wear behavior is affected by the additives in the polymeric coatings. Specifically, PTFE coatings blended with pyrrolidone showed the best friction performance with a friction coefficient of 0.1. Pyrrolidone, usually referred to as poly (vinyl pyrrolidone), has been investigated for medical applications such as articular cartilage replacement due to its excellent low frictional properties [28]. Coatings blended with MoS_2 (PTFE/MoS_2 and Resin/PTFE/MoS_2) exhibited higher wear resistance, as MoS_2 offers favorable surface properties due to its lamellar structure (which can sustain high normal loads and at the same time low shear strength between its planes, resulting in a lubricious low friction surface [7]). The PEEK coating blended with ceramics exhibited the highest friction coefficient and wear rate among the seven coatings tested under dry-oscillatory conditions. Even though PEEK is usually known to be harder than PTFE, and thus expected to have better wear performance, this is not the case in our experiments, where polymeric coatings were tested (vs. bulk). This performance variation of PTFE and PEEK polymers, depending on their form, either bulk or coating, is further discussed in conjunction to unidirectional testing results.

Unidirectional Testing Simulating Swash Plate Compressors

Detailed Experiments Results

The aforementioned seven polymeric coatings are now evaluated using dry-unidirectional testing to better understand their behavior under simulated swash plate compressor conditions. Fig. 8(a) and (b) show the *in-situ* measurements of the friction coefficient and near contact temperature, respectively. Lower friction coefficient values in the range of 0.04–0.1 were observed under unidirectional conditions, while their near contact temperature was 3 to 4 times higher compared to the oscillatory experiments (likely due to an order of magnitude higher sliding speed). The near contact temperature behavior for all the coatings was similar in that at the start of the tests, it was around 22 °C and reached 100 °C immediately after test initiation. After that, depending on the coating, it either remained steady under 150 °C or reached values up to 250 °C. Based on the overall COF and NCT behaviors observed in Fig. 8, coatings could be classified in three groups; (1) the best performing group (which includes PTFE/Pyrrolidone-1, PTFE/Pyrrolidone-2, and PTFE/MoS$_2$) showing the lowest COF around 0.04–0.05 and steady-state NCT of 150 °C, (2) the second group (PEEK/PTFE and PEEK/Ceramic) showing COF around 0.08–0.09 with 200 °C–250 °C NCT, and (3) the scuffed coating group (Resin/PTFE/MoS$_2$, Fluorocarbon) showing abrupt increase of COF and NCT at around 23 min. Interestingly, this classification corresponded to the matrix the coating was made out of, in that PTFE-based coatings were the best performing group, PEEK-based coatings the second best group, and the others the worst performing group. It can be noticed from Fig. 8(a) that the friction coefficient of PTFE-based coatings such as PTFE/Pyrrolidone-2 and PTFE/MoS$_2$ is not only low, but their values are also stable with very small deviations during the whole test, compared to fluctuating friction coefficient of PEEK-based coatings. This fluctuation of the COF is related to friction-induced wear mode, which is seen in the wear track profiles, shown next.

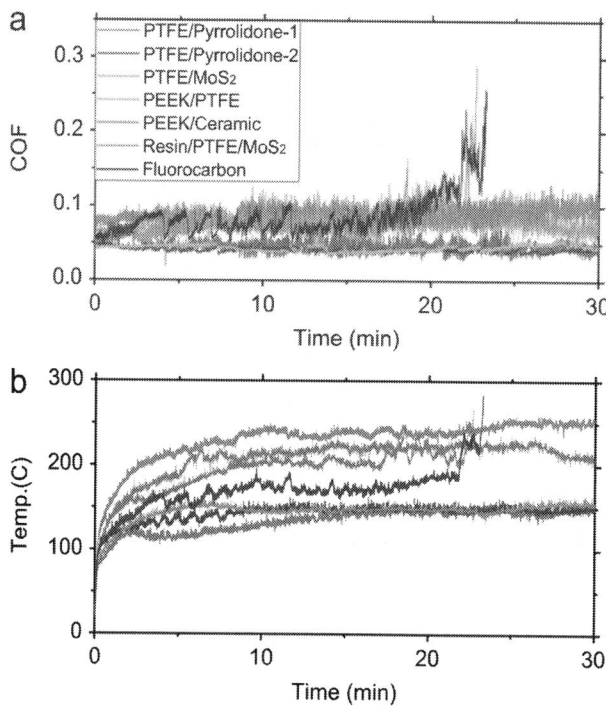

Figure 8: *In-situ* (a) friction coefficient (COF) and (b) near contact temperature (NCT) of seven polymeric coatings during 30 min unidirectional testing.

Fig. 9 shows the wear track profiles of all seven polymeric coatings after 30 min testing under dry-unidirectional conditions. PEEK-based coatings that exhibited fluctuating friction coefficient also show material removal, due to friction-induced wear mode. On the other hand, PTFE-based coatings experienced only mild burnishing of less than 5 μm deep. Fig. 9((a)–(d)) show line scans of two different wear track locations for PTFE/Pyrrolidone-1 and PTFE/Pyrrolidone-2 coating, respectively. Both coatings exhibited similar mild wear behavior with an average wear depth of 2 μm[-4] μm, which was significantly lower compared to oscillatory testing results. The PTFE/MoS$_2$ coating showed slightly higher wear resistance with less than 2 μm of wear as seen in

Fig. 9(e). PEEK/PTFE was worn out completely, Fig. 9(f), and its wear profile was the same as the crown shape of the counter-surface shoe. Despite this high wear, the PEEK-based coating survived the total 30 min testing due to the beneficial effect of the generated wear debris. The two scuffed coatings showed over 60 µm of sharp and deep wear scars, which penetrated through the coating and wear the cast iron substrate as well. Note that these unidirectional test results could more clearly differentiate the wear performance between PTFE- and PEEK-based coatings, compared to the oscillatory experiments. This is attributed to an order of magnitude higher sliding distance with unidirectional testing.

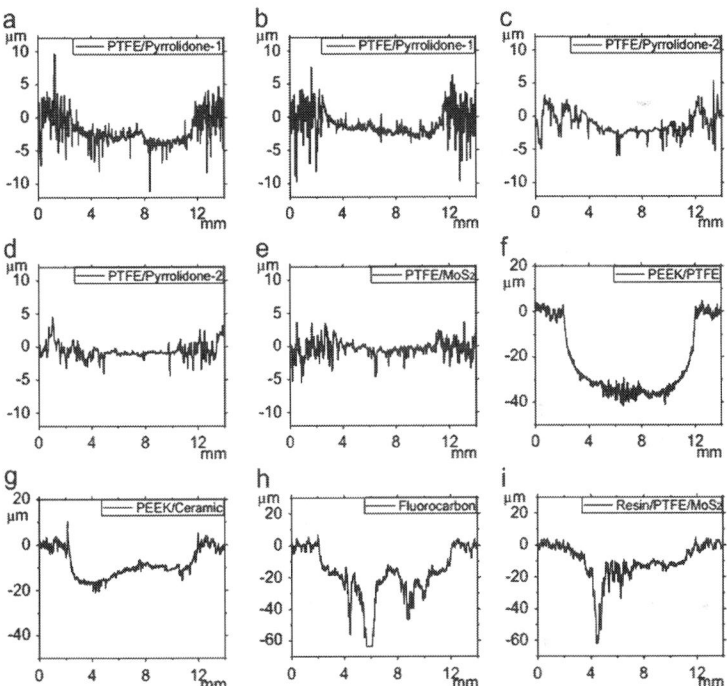

Figure 9: Profilometric wear track measurements of ((a) and (b)) PTFE/Pyrrolidone-1, ((c) and (d)) PTFE/Pyrrolidone-2, (e) PTFE/MoS$_2$, (f) PEEK/PTFE, (g) PEEK/Ceramic, (h) Fluorocarbon, and (i) Resin/PTFE/MoS$_2$ coatings after 30 min unidirectional testing.

In the case of the PTFE-based coatings shown in Fig. 9(a)–(e), which exhibited strong wear resistance, as the shoe was sliding on the coating surface, smoothening of the wear track surfaces is clearly observed. Also, note that the mean plane of the smoothened wear track surface (from 4 mm to 10 mm scan length) is sometimes located slightly higher than the lowest point of the valleys in the original coating surfaces (both edges of wear profiles from 0 to 2 or from 12 mm to 14 mm scan length). This is attributed to the fact that the very fine PTFE-based wear debris (observed after testing) was filling the valleys and pits of the initial rough coating surfaces, and thus, solid polymer lubricant can stay continuously trapped inside the wear track, thus effectively lubricating the dry interface preventing catastrophic failure. None of these behaviors were observed in the other coatings, which alludes to the fact that this is a critical mechanism of polymeric coatings in determining their tribological performance. This "filling effect" of wear debris is also seen in the optical pictures in Fig. 10, showing the wear tracks after unidirectional testing. The wear tracks of PTFE/Pyrrolidone-2 and PTFE/MoS$_2$ coatings are very glossy after sliding because the surface got smoothened due to the filling effect of the wear debris. On the contrary PEEK based coatings generated flake-like debris, and resulted in continuous coating material removal and thus higher wear rates. In the case of the scuffed coatings (Fluorocarbon and Resin/PTFE/MoS$_2$), complete penetration of the coatings was observed, thus exposing the cast iron substrate surface as seen in Fig. 10(d). Because the counterpart (shoe) surface is extremely smooth (0.035 µm), the filling effect is not observable on the shoe surface, and in this case, usually a very thin transfer film is formed [25].

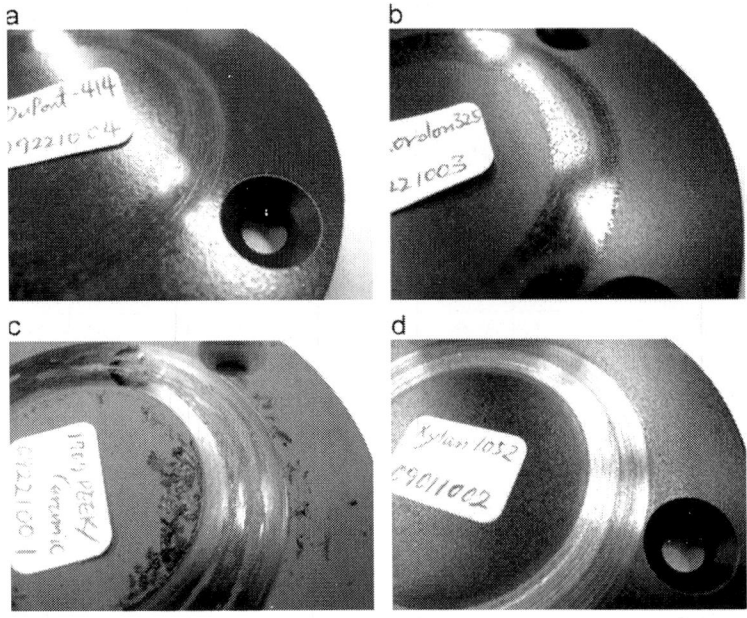

Figure 10: Optical images of the wear track surfaces after 30 min unidirectional testing; (a) PTFE/Pyrrolidone-2, (b) PTFE/MoS2, (c) PEEK/Ceramic, and (d) Resin/PTFE/MoS$_2$.

Oscillatory vs. Unidirectional; Coating vs. Bulk and PTFE vs. PEEK

The wear rates for the unidirectional tests were also calculated and summarized in Table 3, along with average COF and NCT values during the steady-state period from 10 min to 30 min. PTFE/Pyrrolidone-2 exhibited the lowest average COF value of 0.043, PTFE/MoS$_2$ the second lowest, and then PTFE/Pyrrolidone-1. However, their difference was very small with all three coatings being in the range of 0.043–0.047. These are significantly low friction coefficient values, bearing in mind that these are oil-less aggressive operating conditions. PTFE/MoS$_2$ coating showed the lowest wear rate with a value less than 10^{-7} mm^3/N m. The friction coefficient for the PEEK-based coatings was almost twice as high compared to the PTFE-based coatings, and their wear rate was in the range of 6.73×10^{-6} mm^3/N m–1.63×10^{-5} mm^3/N m (i.e., an order of magnitude higher than PTFE-based coatings).

Table 3: Average COF, NCT and wear rates of the polymeric coatings tested under dry-unidirectional conditions for 30 min

	Polymeric coatings	Ave.COF(Std.) (10–30 min)	Wear rate (Std.) (mm³/N m)	NCT (Std.) (°C) (10–30 min)
Group 1	PTFE/ Pyrrolidone-1	0.047 (0.007)	1.54E-6 (2.11E-7)	142.8 (5.9)
	PTFE/ Pyrrolidone-2	0.043 (0.003)	1.23E-6 (2.97E-7)	148.6 (2.3)
	PTFE/MoS₂	0.044 (0.005)	3.76E-7 (8.64E-8)	148.7 (3.8)
Group 2	PEEK/PTFE	0.079 (0.008)	1.63E-5 (1.28E-6)	219.8 (5.2)
	PEEK/Ceramic	0.092 (0.010)	6.73E-7 (9.77E-7)	241.9 (6.5)
Group 3	Resin/PTFE/ MoS₂	Scuffed	9.01E-6 (2.39E-7)	–
	Fluorocarbon	Scuffed	1.52E-5 (6.02E-7)	–

Fig. 11 shows a direct comparison of both friction (linear scale) and wear (logarithmic scale) behavior of all coatings. From this comparison, it is clearly seen that the wear rates for all coatings tested under oscillatory conditions, are clustered in very small range of 2×10^{-5} mm³/N m–5×10^{-5} mm³/N m, whereas, in the case of unidirectional testing conditions, the wear rate values vary in the range of 10^{-7} mm³/N m–10^{-5} mm³/N m. Also, in addition to the lower wear rates, the coatings tested under unidirectional conditions, exhibited lower friction coefficient values, compared to oscillatory testing conditions. It is cautioned against generalizing such finding, because the testing parameters (sliding speed, contact pressure and the shape of pin) were different for each testing condition, simulating specific type of compressor conditions. FromFig. 11(a), PTFE/MoS₂ coating located in the lower left hand corner of the plot is the best performing coating for swash plate compressor simulation, whereas, for piston-type compressors, PTFE/Pyrrolidone-1 seems to be more favorable, as seen in Fig. 11(b).

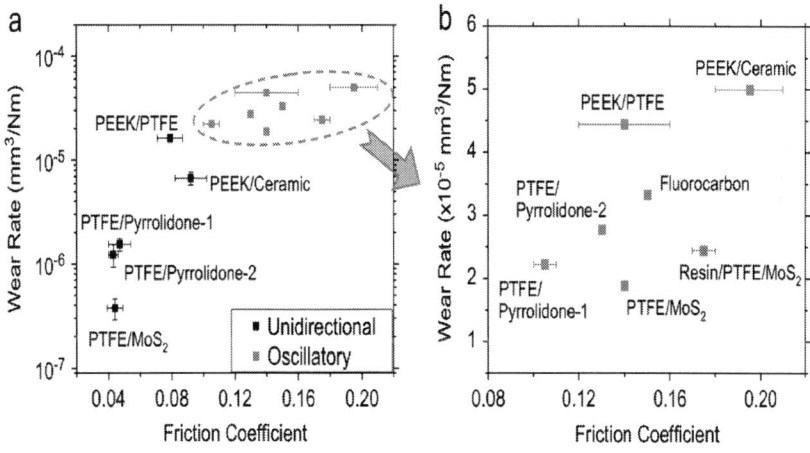

Figure 11: (a) Friction coefficient vs. wear rate of seven polymeric coatings under both unidirectional and oscillatory testing and (b) zoom-in of only oscillatory testing results.

A general finding from this work is that PTFE-based coatings exhibited better tribological performance than PEEK-based coatings, under both oscillatory and unidirectional conditions, and, interestingly, this was not the case for bulk polymers. Cannaday and Polycarpou [14], for example, demonstrated that bulk form of PEEK blended with graphite exhibited the lowest friction coefficient and wear rate compared to PTFE blended with graphite in both R-134A and ambient air environments (under unidirectional sliding conditions). Also, it was shown in Ref. [29] that bulk form of PEEK blended with PTFE exhibited lower wear rate than any other PTFE blends, such as PTFE/PEEK, PTFE/CNT and PTFE/Si$_3$N$_4$. These performance variations of PTFE and PEEK polymers depending on their form, either bulk or coating, could be attributed to the different wear mechanism for each case. For bulk materials, continuous scratching and peeling of polymers are the dominant wearing mechanism, and thus making harder materials tend to exhibit lower wear rate. This is why the harder PEEK blends usually perform better than PTFE blends when they are used in bulk form. In the case of coatings, wear debris comes to effectively serve as solid lubricant at some point due to usually very hard substrate material (cast iron in this work) which is not the case in bulk polymers where continuous material removal occurs. Therefore, the filling effect of the wear debris explained in Fig. 9 becomes dominant in the wear behavior of polymeric coatings. In this case, the structure of the coating play a key role in determining the formation of wear debris and the overall behavior of the coating. Originally, both PTFE and PEEK polymers are classified as crystalline thermoplastic along with the most widely used low density polyethylene (LDPE) [30]. However, the sputtered PTFE films usually have an amorphous structure, resulting in very fine wear debris [7] which is more favorable for effective lubrication. Obtaining an amorphous PEEK is almost impossible due to its high crystallizing speed and low thermal conductivity [20].

SEM images of the worn surfaces for PEEK/PTFE and PTFE/Pyrrolidone-2 coatings (shown in Fig. 12) depict the above mentioned structural differences between the two coatings, as well as the wear debris formation for each coating. From Fig. 12(a), it can be clearly seen that a big "chunk" of coating material was removed from the PEEK/PTFE coating surface by sliding, which explains the large size of wear debris for this coating. Also, the sub-surface area as marked in the figure shows the porous structure of the PEEK/PTFE coating. However, in the case of PTFE/Pyrrolidone-2 coating (Fig. 12(b)) only mild plastic plowing was observed without any signs of severe material removal from the surface. Also, it appears that fine powders were compressed together showing an extensively smooth top surface of PTFE/Pyrrolidone-2 coating, and thus resulting in its uniform microstructure. This is the reason the wear debris of PEEK-based coatings was larger and more flake-like as seen in Fig. 10, and thus could not effectively fill the rough surfaces, resulting in higher friction coefficient and wear rate, compared to PTFE-based coatings. Thus, in the form of coatings, the wear performance of PTFE blends is improved, compared to the PEEK blends.

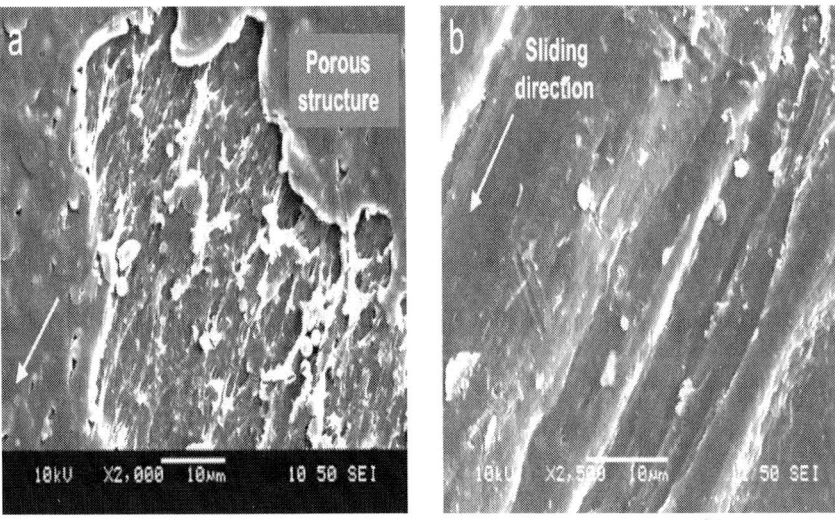

Figure 12: SEM images of the worn surfaces of (a) PEEK/PTFE coating and (b) PTFE/Pyrrolidone-2 coating after 30 min dry unidirectional sliding test.

Durability Unidirectional Testing

Since the 30 min tests (corresponding to 6.75 km sliding distance) might not be sufficiently long to validate the polymeric coating's reliability in compressor applications, and to also further substantiate the superiority of the PTFE-based coatings, 3 h long duration unidirectional tests (corresponding to 40.5 km sliding distance) were performed for PTFE/Pyrrolidone-2, PTFE/MoS$_2$, and PEEK/PTFE coatings, and shown in Fig. 13. Even though none of these coatings showed scuffing with the 30 min tests, PEEK/PTFE coating eventually scuffed after 105 min, with sharp increase of the friction coefficient as seen in Fig. 13(a). Both PTFE-based coatings exhibited very stable and lower friction coefficient than the PEEK-based coatings, however, the COF of PTFE/MoS$_2$ coating started to fluctuate after 2 h and showed somewhat unstable behavior with increased values up to 0.13. Even though PTFE/MoS$_2$ coating did not scuff until the end of the 3 h testing, it would have eventually scuffed. This is because not only its COF became unstable, but its near contact temperature was also gradually increasing and reached 200 °C. The best performance was observed with PTFE/Pyrrolidone-2 coating with very stable friction coefficient, less than 0.05 during the whole test, and its NCT also being very stable and less than 150 °C, even after 3 h.

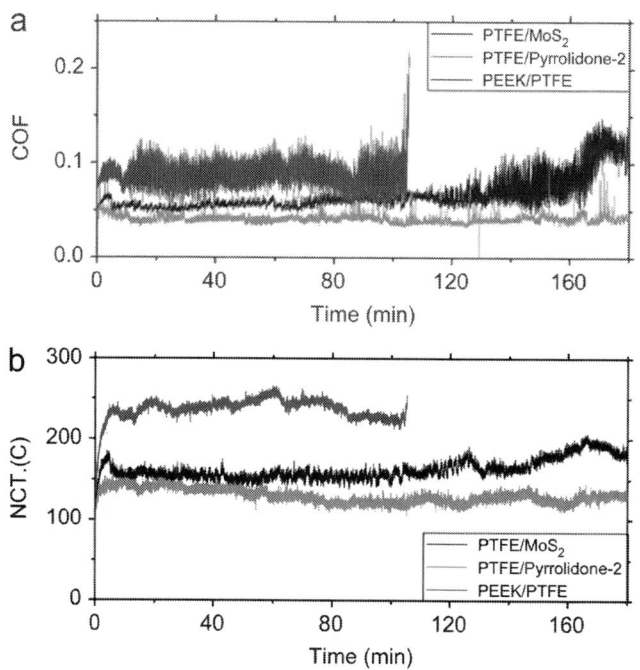

Figure 13: *In-situ* (a) friction coefficient and (b) near contact temperature of PTFE/MoS$_2$, PTFE/Pyrrolidone-2, and PEEK/PTFE coatings during 3 hour uni-directional testing.

The wear track profiles of the two PTFE-based coatings after the durability 3 h tests are shown in Fig. 14((c) and (d)) and compared with the 30 min test results, which are repeated in Fig. 14((a) and (b)). Both coatings exhibited only mild burnishing with less than 5 μm of wear depth after 30 min test. After the 3 h tests, more material removal occurred with the PTFE/MoS$_2$ coating surface showing over 20 μm of wear depth, whereas, in the case of PTFE/Pyrrolidone-2 coating, surprisingly, still only 5 μm depth of material removal was observed. As discussed earlier, after 30 min testing, PTFE/MoS$_2$ coating seemed to perform the best under unidirectional sliding conditions. However, as shown after the 3 h tests, PTFE/Pyrrolidone-2 coating had superior wear performance to the other coatings, relevant to compressor applications.

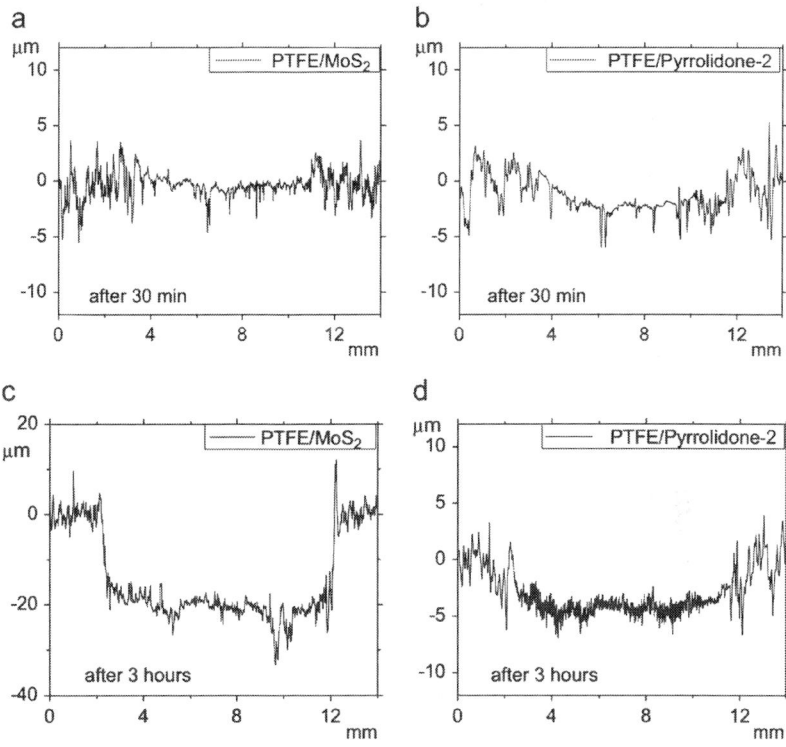

Figure 14: Profilometric wear track measurements of (a) PTFE/MoS$_2$ and (b) PTFE/Pyrrolidone-2 after 30 min, and (c) PTFE/MoS$_2$ and (d) PTFE/Pyrrolidone-2 after 3 h under dry-unidirectional testing.

CONCLUSIONS

Different polymeric coatings based on PTFE, PEEK, fluorocarbon and resin were tested using a specialized tribometer under aggressive conditions simulating critical tribopairs of both piston-type (reciprocating motion) and swash plate (unidirectional motion) air-conditioning and refrigeration compressors. The following conclusions could be drawn:

- Higher friction coefficient values in the range of 0.1–0.2 and wear rate values of 10^{-5} mm^3/N m were measured under oscillatory conditions. Better performance with lower friction coefficient values of 0.04–0.1 and wear rates of 10^{-7} mm^3/N m were measured under unidirectional conditions. Under both operating conditions (but more so for unidirectional), polymeric coatings exhibited acceptable to superior tribological performance, under aggressive oil-less compressor conditions, and thus are likely viable candidates for the next generation of oil-less compressors.

- PTFE-based coatings (PTFE/Pyrrolidone and PTFE/MoS$_2$) generally performed better than PEEK-based coatings (PEEK/PTFE and PEEK/Ceramic) under both piston-type (oscillatory motion) and swash plate (unidirectional motion) compressor conditions, which was not the case for bulk polymer blends (where typically PEEK composites filled with PTFE performed best).

- The effect of polymer coating wear debris, serving as a solid lubricant on the hard substrate surface, was shown to be more dominant in determining the overall wear behavior of polymeric coatings than the mechanical properties of the polymer coating itself. The structural changes in the PTFE coatings from semi-crystalline to amorphous caused by the coating process, resulted in very fine wear debris enhancing its lubricity. PEEK-based polymer coatings were still exhibiting crystalline structure after the coating process, due to its high crystallizing speed, thus resulting in large and flake-like wear debris.

- Durability or time-to-failure (3-h duration) unidirectional testing corresponding to 40.5 km sliding distance showed superb friction and wear behavior of PTFE-based coatings, especially PTFE/Pyrrolidone-2, thus demonstrating its potential applicability for use in oil-less compressors.

ACKNOWLEDGMENTS

This research work was supported by the 25 member companies of the Air Conditioning and Refrigeration Center (ACRC), an Industry-University Cooperative Research Center at the University of Illinois at Urbana-Champaign.

REFERENCES

1. S.R. Pergande, A.A. Polycarpou, T.F. Conry, Nanomechanical properties of aluminum 390-T6 rough surfaces undergoing tribological testing, Journal of Tribology 126 (2004) 573–582.

2. T.A. Solzak, A.A. Polycarpou, Tribology of WC/C coatings for use in oil-less piston-type compressors, Surface and Coatings Technology 201 (2006) 4260–4265.

3. N.G. Demas, A.A. Polycarpou, Ultra high pressure tribometer for testing CO2 refrigerant at chamber pressures up to 2000 psi to simulate compressor conditions, Tribology Transactions 49 (2006) 291–296.

4. Y.Z. Lee, S.D. Oh, Friction and wear of the rotary compressor vane-roller surfaces for several sliding conditions, Wear 255 (2003) 1168–1173.

5. A. Bloyce, Coatings for high performance pumps and compressors, World Pumps 2000 (400) (2000) 43–45.

6. A.Y. Suh, A.A. Polycarpou, Analytical determination of the surface energy of sub-5 nm head-disk interfaces accounting for multilayer effects, Journal of Applied Physics 99 (2006) 08N111.

7. K. Holmberg, A. Matthews, Coatings tribology: properties, mechanisms, in: B.J. Briscoe (Ed.), Techniques and Applications in Surface Engineering, second ed.,Elsevier B.V., Oxford, UK, 2009.

8. T.A. Solzak, A.A. Polycarpou, In: Proceedings of the Ninth Biennial ASME Conference on Engineering Systems Design and Analysis, ESDA2008-59380, Haifa, Israel (2008).

9. Q. Zhao, Y. Liu, H. Muller-Steinhagen, G. Liu, Graded Ni-P-PTFE coatings and ¨ their potential applications, Surface and Coatings Technology 155 (2002) 279–284.

10. H.C. Sung, Tribological characteristics of various surface coatings for rotary compressor vane, Wear 221 (1998) 77–85.

11. R.L. Fusaro, Self-lubricating polymer composites and polymer transfer film lubrication for space applications, Tribology International 23 (1990) 105–122.

12. M. Yamane, T.A. Stolarski, S. Tobe, Wear and friction mechanism of PTFE reservoirs embedded into thermal sprayed metallic coatings, Wear 263 (2007) 1364–1374.

13. N.P. Suh, Tribophysics, Prentice-Hall, Englewood Cliffs, NJ, 1986.

14. M.L. Cannaday, A.A. Polycarpou, Tribology of unfilled and filled polymeric surfaces in refrigerant environment for compressor applications, Tribology Letters 19 (4) (2005) 249–262.

15. W. Hufenbach, K. Kunze, J. Bijwe, Sliding wear behaviour of PEEK-PTFE blends, Journal of Synthetic Lubrication 20 (3) (2003) 227–240.

16. B. Lal, S. Alam, G.N. Mathur, Tribo-investigation on PTFE lubricated PEEK in harsh operating conditions, Tribology Letters 25 (1) (2007) 71–77.

17. T.A. Stolarski, Tribology of polyetheretherketone, Wear 158 (1992) 71–78.

18. Z.P. Lu, K. Friedrich, On sliding friction and wear of PEEK and its composites, Wear 181-183 (1995) 624–631.

19. T.C. Stening, C.P. Smith, P.J. Kimber, Polyaryletherketone: high performance in a new thermoplastic, Modern Plastics International 8 (1982) 54–57, March.

20. G. Zhang, W.-Y. Li, M. Cherigui, C. Zhang, H. Liao, J.-M. Bordes, C. Coddet, Structures and tribological performances of PEEK-based coatings designed for tribological application, Progress in Organic Coatings 60 (2007) 39–44.

21. G. Zhang, H. Liao, H. Li, C. Mateus, J.-M. Bordes, C. Coddet, On dry sliding friction and wear behaviour of PEEK and PEEK/SiC-composite coatings, Wear 260 (2006) 594–600.

22. N.L. McCook, D.L. Burris, G.R. Bourne, J. Steffens, J.R. Hanrahan, W.G. Sawyer, Wear resistant solid lubricant coating made from PTFE and epoxy, Tribology Letters 18 (1) (2005) 119–124.

23. N.G. Demas, A.A. Polycarpou, Tribological performance of PTFE-based coatings for air-conditioning compressors, Surface and Coatings Technology 203 (2008) 307–316.

24. D. Dascalescu, K. Polychronopoulou, A.A. Polycarpou, The significance of tribochemistry on the performance of PTFE-based coatings in CO_2 refrigerant environment, Surface and Coatings Technology 204 (2009) 319–329.

25. E. Escobar Nunez, S. Yeo, K. Polychronopoulou, A.A. Polycarpou, Tribological study of high bearing blended polymer-based

coatings for air-conditioning and refrigeration compressors, Surface and Coatings Technology 205 (2011) 2994–3005.

26. J. Zhang, A.A. Polycarpou, J. Economy, An improved tribological polymer coating system for metal surfaces, Tribology Letters 38 (3) (2010) 355–365.

27. T.A. Solzak, A.A. Polycarpou, Tribology of hard protective coatings under realistic operating conditions for use in oilless piston-type and swash-plate compressors, Tribology Transactions 53 (3) (2010) 319–328.

28. J.K. Katta, M. Marcolongo, A. Lowman, K.A. Mansmann, Friction and wear behavior of poly(vinyl alcohol)/poy(vinyl pyrrolidone) hydrogels for articular cartilage replacement, Journal of Biomedical Materials Research Part A 83A (2) (2007) 471–479.

29. D.L. Burris, W.G. Sawyer, Tribological behavior of PEEK components with compositionally graded PEEK/PTFE surfaces, Wear 262 (2007) 220–224.

30. H.F. Brinson, L.C. Brinson, Polymer engineering science and viscoelasticity, an introduction, Springer, New York, NY, 2008.

Chapter 8

Turbomachinery Component Manufacture by Application of Electrochemical, Electro-Physical and Photonic Processes

Fritz Klocke[a], Andreas Klink[a,] Drazen Veselovac[a],
David Keith Aspinwall[b], Sein Leung Soo[b], Michael
Schmidt[c], Johannes Schilp[d], Gideon Levy[e], and
Jean-Pierre Kruth[f]

[a]Laboratory for Machine Tools and Production Engineering, WZL,
RWTH Aachen University, Aachen, Germany

[b]Machining Research Group, School of Mechanical Engineering,
University of Birmingham, Birmingham, United Kingdom

[c]Bayerisches Laserzentrum GmbH, BLZ, Erlangen, Germany

[d]Institute for Machine Tools and Industrial Management IWB,
Technische Universität München, München, Germany

[e]Centro Para o Desenvolvimento Rapido e Sustentado de Produto
CDRSP, Instituto Politecnica de Leiria IPL, Leiria, Portugal

[f]Division PMA (Production Engineering), University of Leuven (KU
Leuven), Leuven, Belgium

ABSTRACT

This paper presents an overview of the current technological and economical capabilities of electrochemical (ECM-based), electro-physical (EDM-based) and photonic (Laser-/EBM-based) additive and removal processes for turbomachinery component manufacture. Starting with the industrial demands and challenges of today, the technologies are reviewed in detail regarding achievable geometrical precision and surface integrity as well as material removal and deposition rates for conventionally difficult-to-cut Ti- and Ni-based alloys and dedicated steels. Past, existing and future areas of technology application of these advanced non-mechanical manufacturing processes are discussed. The paper focusses on the description of shaping processes therefore excludes pure welding or coating applications.

INTRODUCTION

The demand for turbomachinery systems such as aero-engines, stationary gas and steam turbines as well as turbochargers for engines is constantly growing due to the increasing worldwide requirement for energy and mobility. In contrast, conventional energy resources such as oil, gas and coal together with important raw materials are shrinking while environmental pollution due to CO_2 and NOx emissions is on the rise. Thus, energy and fuel prices as well as costs for environmental protection and sustainability are constantly increasing, necessitating the development and introduction of highly efficient turbomachinery systems.

Taking the aerospace sector as an example, air traffic is resiliently growing at a rate of 4–5% a year both for revenue passenger (RPK) as well as cargo traffic tonne kilometres (RTK), practically doubling within 15 years. According to the 'Global Market Forecast 2012-32', Airbus predicts a doubling of the passenger aircraft fleet ($\geq$ 100 seats: single/twin-aisle and very large) from 16,094 to 33,651 by 2032. Including replacements, some 28,355 new aircraft deliveries are anticipated. Similar numbers are presented in Boeing's 'Current Market Outlook 2013-32' showing the 20,310 aircraft (regional jets, single aisle, small/medium/large widebody) currently in service increasing to 41,240 by 2032 with new deliveries of 35,280 [29], [86] and [123]. In terms of aeroengines, Rolls-Royce expects ~68,000 deliveries (including business jets) over the period 2012-31, with a market value of $975 billion [164]. Adding to this, the servicing of commercial engines involving maintenance, repair and overhaul (MRO) is also growing in importance. Within GE Aviation, the service market for 2011 amounted to $7.2 billion while the new engine market was $4.9 billion [100].

Besides market growth, the challenges faced by industry are also growing, because future aircraft including the engines must also be more fuel efficient, quieter and cleaner due to official regulations and agreements. The new ACARE (Advisory Council for Aviation Research and Innovation in the EU) goals for 2050, schedule a reduction of 75% in CO_2, 90% in NOx and 65% in noise relative to 2000 [2] and [86]. In summary, there is an extensive and pressing need for design – as well as advanced manufacturing and repair technologies able to handle the current and growing future demands for turbomachinery components.

CHALLENGES OF TURBOMACHINERY COMPONENT MANUFACTURE

Core functional components of turbomachinery systems are characterised by the use of dedicated high temperature, high specific strength and wear-resistant materials (Fig. 1).

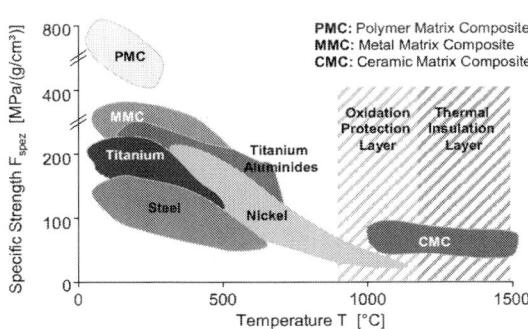

Figure. 1: Specific strength of materials as a function of working temperature for turbo machinery components. Based on [181].

Machining such "difficult-to-cut" materials using conventional means is very challenging, often resulting in low material removal rates (MRR), reduced precision due to high cutting forces, high tooling costs due to increased wear and consequently low process efficiency [178]. In addition, the resulting surface integrity is often characterised by thermo-mechanically altered or even damaged rim zones [97], [178] and [196]. Thus, the utilisation of technological as well as economically suitable manufacturing technologies is of great interest.

Taking the aerospace sector again as an example, Fig. 2 shows the specific areas of application for preferred Ti and Ni-based alloys in aeroengines. The temperature capability of such materials is constantly increasing through the development of new materials with different primary manufacturing technologies[158] (Fig. 3). Due to the absence of grain boundaries, single crystal materials exhibit far better creep properties than polycrystalline materials and can therefore be utilised at higher temperatures [34]. The use of such new materials and especially the advanced gamma titanium aluminides (for compressor as well as turbine applications) [13] and polymer matrix composites – PMC (for fan blading components), (Fig. 2), require amongst others, the development of appropriate manufacturing technologies.

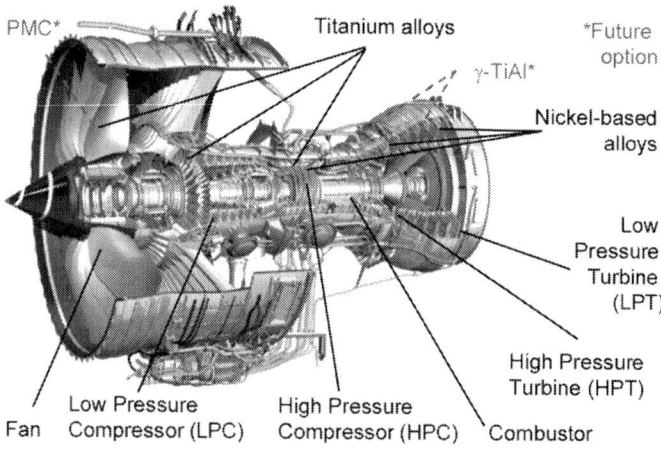

Figure. 2: Current and future temperature specific application of materials in aero engines (example: Rolls Royce Trend 800 engine).

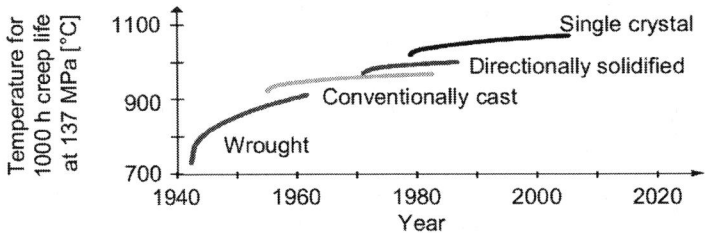

Figure. 3: Evolution of high-temperature strength capability of turbine blading nickel-based super alloys. Based on [158].

The most important turbomachinery steel, Ti- and Ni-based materials discussed in this paper are:

- *Steel alloys*: X22CrMoV211; X12CrNiWTiB16-13
- *Ti-based alloys*: Ti–6Al–4V (Ti64/Ti6Al4V); Ti–6Al–2Sn–4Zr–2Mo–0.1Si (Ti6242); Ti6246; Ti–5Al–2Sn–2Zr–4Mo–4Cr (Ti17)
- Gamma titanium aluminides (γ-TiAl): TNM; GE 48-2-2; 45 XD
- *Ni-based superalloys*: Inconel 718** (In718); In718 DA***; IN100**; Inconel 738**; Inconel 939**; MAR-M002°; MAR-M247°; Waspalloy*; Udimet 720*; Nimonic 105*; Nimonic 713*; Rene88**; RR1000°*; CMSX4°°; LEK94°°

Legend:wrought, cast, directionally solidified, single crystal powder metallurgical, direct aged

In order to increase aeroengine economic and ecological efficiency, current focus centres on the enhancement of propulsive as well as thermal effectiveness [31], [98] and [198]. Propulsive efficiency can mainly be improved by realisation of higher by-pass ratios such as via the concept of Geared Turbo Fans[81]. Due to limited ground clearance of aircraft (classical design) and therefore limited fan diameters, the core engine has subsequently also to be reduced in size. The thermal efficiency can be further increased through higher temperature combustion requiring new high temperature resistant and lightweight materials (e.g. graded materials, single crystal, metal matrix composites – MMC and ceramic metal composites – CMC) and better cooling concepts (new cooling hole geometries, double walled bladings) as well as thermal barrier coatings (TBC) [32], [64], [175], [180] and [181]. As a consequence, the need for more Ni-based compressor/turbine stages can be expected. Additionally, improved aero-dynamic and lightweight construction designs involving "hyper-polished" airfoils, elliptical leading/trailing edges or blisk manufacture, (Fig. 4), can further increase stage pressure ratios and thus efficiency [36] and [38].

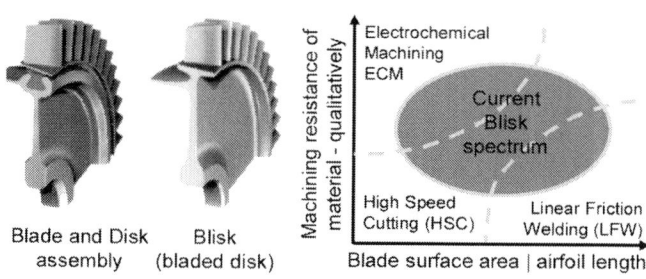

Figure. 4: Blisk principle design [163] and exemplarily, qualitative classification of currently used machining technologies as a function of blade surface are size (and airfoil length) and material machinability, [36].

In order to overcome current manufacturing limitations of conventional machining and to extend the potential of Design For Manufacture (DFM), an evaluation of the capabilities of advanced, non-mechanical, single process technologies, as well as new process chains both for initial manufacture and repair is necessary [37]. Comparing e.g. milling and ECM (Fig. 5), the MRR is reduced hyperbolical with the cutting tool overhang while constantly increasing with the ECM working area for comparable axis scales during blisk slot roughing. Besides productivity, the criticality of aeroengine failures necessitates appropriate work piece surface integrities [196]. Also, economic capabilities have to be evaluated against the background of constantly growing volumes in turbomachinery serial production [22].

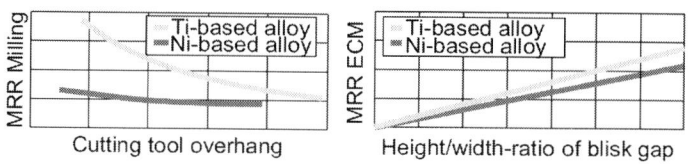

Figure.5: Conceptual qualitative comparison of technological potential of conventional milling and ECM for machining of blisk blading gaps.

The paper reviews comprehensively the past, current and future technological and economical capabilities of ECM, EDM and photonic (laser/EBM) additive and subtractive processes for turbomachinery component manufacture. It focusses on the description of shaping processes excluding pure welding or coating applications. As the machining of composite materials (PMC, MMC, CMC) with electrically non-conductive phases cannot be executed using ECM/EDM and is generally inefficient with lasers, the paper focusses on the machinability analysis of steel, Ti and Ni-based alloys.

TECHNOLOGICAL CAPABILITIES OF ADVANCED PROCESSES

Electrochemical Machining (ECM)

Introduction

The major advantages of ECM are its process specific characteristics of high material removal rate in combination with almost no tool wear. Due to cost intensive tool pre-developing processes and rather high investment outlays for the machine tools however, ECM is specifically used in large batch size production and represents an alternative manufacturing technology for turbomachinery components. In addition, high material removal rates can be realised while achieving good workpiece surface quality without the occurrence of white layers, heat affected zones or strain hardening [104], [110] and [112]. This section gives a specific overview on the technological capabilities for the production of different geometrical features of turbomachinery components.

Material Specific Removal Rates and Resulting Surface Integrity

Independent of the type of ECM process – direct current (DC) or electrically and mechanically pulsed processes (PECM) – material removal is only dependent on the electrochemistry of the workpiece materials in combination with the electrolyte and the programmed current densities. For ECM tool design, it is necessary to know the local gap details during the process. The local gap width can initially be calculated with a combination of Ohm's and Faraday's law (for complex shaped geometries and long flow lengths the approximation loses its validity). Eq. (1) shows the effective material removal rate V_{eff} based on the specific removal rate $V_{sp,alloy}$ according to Faraday's law reduced by the current efficiency η [104]. In the pure form of Faraday's law, only one electrochemical valency is considered for each element and the equation has to be corrected by η, which also can be considered as an efficiency factor. Furthermore, the calculation of local effective gap s_{eff} by the combination of Ohm's and Faraday's law is only applicable in the frontal gap[104]. For complex shaped geometries, it can roughly be corrected by the angle of inclination of the workpiece contour α (Eq. (2)), [112].

$$V_{eff} = V_{sp,alloy} \cdot \eta = \frac{\eta}{\rho_{alloy}} \cdot \sum_{i=1}^{n} \frac{w_i}{100} \cdot \frac{M_i}{z_i \cdot F}$$

(1)

$$S_{eff,\alpha} = \frac{(U - \Delta U) \cdot V_{eff} \cdot \kappa}{\upsilon_f \cdot \sin\alpha} = \frac{(U - \Delta U) \cdot V_{eff} \cdot \kappa}{\upsilon_f \cdot \sqrt{1 - \cos^2\alpha}}$$

(2)

In practice, it is necessary to experimentally determine the effective material removal rate by analysing the behaviour of feed rate as a function of current density. Typical results are shown in Fig. 6 for the DC-machining of cylindrical holes with a tool diameter of 6 mm and internal flushing.

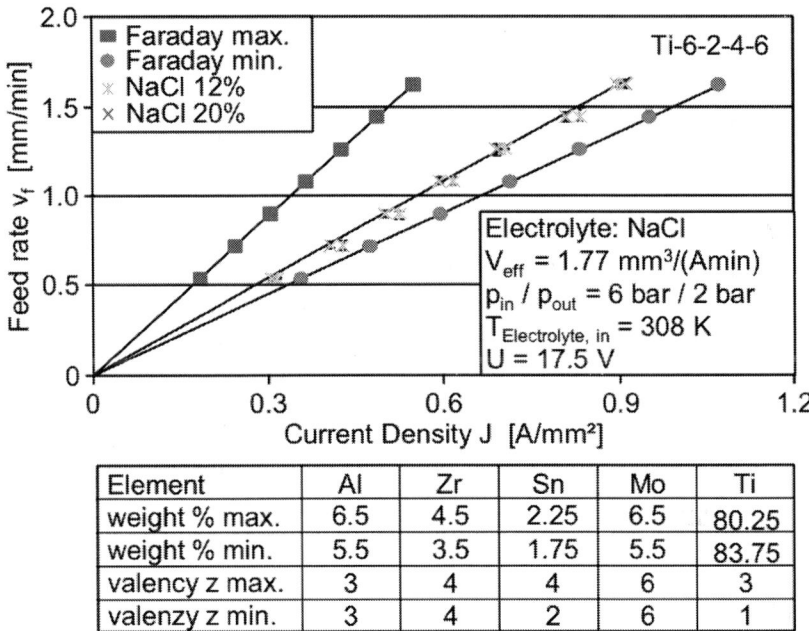

Element	Al	Zr	Sn	Mo	Ti
weight % max.	6.5	4.5	2.25	6.5	80.25
weight % min.	5.5	3.5	1.75	5.5	83.75
valency z max.	3	4	4	6	3
valenzy z min.	3	4	2	6	1

Figure.6: Experimental result of feed rate depending on current density of Ti-6-2-4-6 in comparison to Farady's law, [112].

The linear behaviour and zero dependence of electrolyte concentration on effective material removal rate are typical of ECM processes at high current densities. Fig. 7 summarises the averaged effective material removal rates of relevant turbomachinery alloys. The linear curves were combined into a single function denoted as $V_{eff,\emptyset}$, which is the averaged effective material removal rate. All of the titanium alloys had a near identical $V_{eff,\emptyset}$ of ~1.78 mm³/(A min). In the case of nickel-based alloys, finer grained microstructures lead to better electrochemical machinability and faster dissolution. Consequently, superalloys manufactured via powder metallurgy (PM) techniques show the best electrochemical machinability. In general, nickel-based alloys dissolve faster than titanium alloys [112].

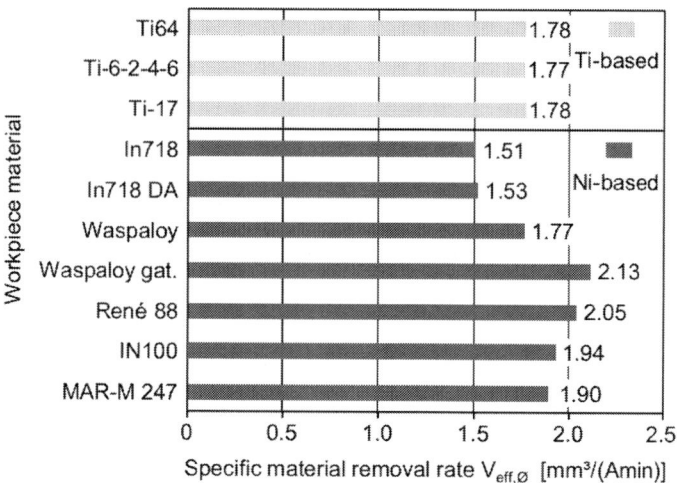

Figure. 7: Averaged effective material removal rates for different titanium- and nickel-based alloys, [112].

Besides the proportional interrelationship between feed rate v_f (equal to the resulting dissolution velocity v_A) and current density, hyperbolic interrelationships can be identified for the frontal gap size s_{90} ($s_{eff}, \alpha; \alpha = 90°$) as well as the resulting surface roughness (Ra) and the current density [110]. Therefore, high material removal rates can in principle be achieved simultaneously with small frontal gap sizes which give the highest precision while also producing low surface roughness for surface smoothing and polishing effects. Fig. 8 shows results for the machining of Ti–6Al–4V and Inconel 718.

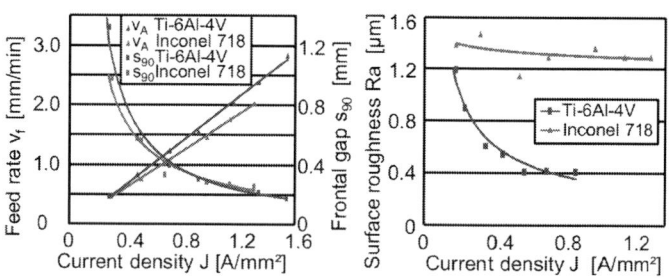

Figure. 8: Machining characteristics – feed rate, frontal gap size and resulting surface roughness depending on current density, [110].

The fact that the hyperbolic response of Inconel 718 is not as pronounced as Ti–6Al–4V is a result of the specific microstructure of the nickel-based alloy employed. Such curve relationships mean that higher feed rates can be used to produce advanced surface quality at the front edge. Conversely, the surface quality at the lateral area of 3D-structures declines with low current densities [110].

The resulting microstructure of the part surfaces after ECM strongly depends on the specific composition of the bulk material. Although no thermally damaged or mechanically deformed rim zones occur with ECM, different material phases or crystallographic orientations of the alloys have different electrochemical dissolution behaviours (up to inter-crystalline and pitting corrosion) resulting in different removal speeds[179]. Depending on the size of the phases and different heat treatment – a defined roughness or waviness of the surface can result following ECM.

Fig. 9 gives an overview on surface integrities of different materials after specific electrochemical treatments. For Inconel 718 and Inconel 718 DA which has a more fine-grained microstructure, a smooth and flat surface finish without any rim zone can be seen in the cross sections when employing optimised ECM parameters. For Ti–6Al–4V, a slightly faster dissolution of the α-phase is visible in the detailed cross section with higher magnification showing a slight waviness – but without any rim zone – of the surface. For γ-TiAl machining, overlaying α_2-lamellas with thicknesses of <2 µm (due to a much reduced dissolution velocity [21]) results in high roughness after ECM shown in the detailed top view of the surface.

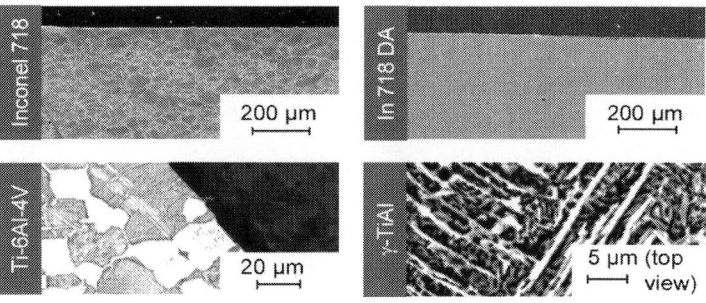

Figure.9: Achievable surface integrity for different titanium and nickel-based alloys after ECM machining. Based on [21] and [112].

New Process Modelling Approaches

The main reasons for high tooling costs with ECM are the reliance on only knowledge-based, iterative cathode designing process. After a test run the workpiece has to be measured and the difference between target and actual geometry is subtracted from the cathode and so forth, yet the theoretical background of the ECM process with all its different physical aspects, cf. Fig. 10, is well known [30] and [109].

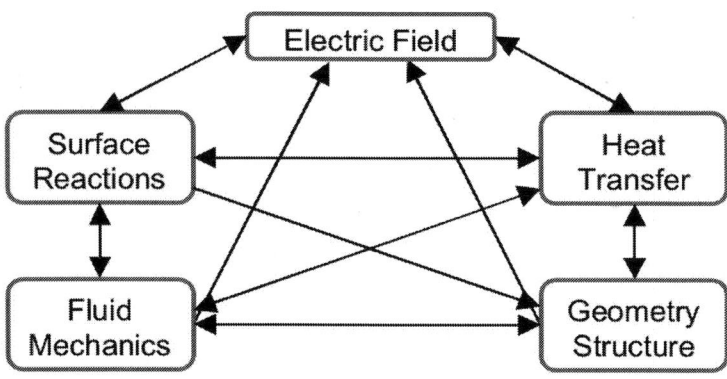

Figure.10: Various physics coupling in modelling of ECM processes. Based on [110] and [194].

Improvements in computing power and simulation tools allow "multi-physics" approaches to simulate the complex ECM processes in one comprehensive approach. Thus, it will be possible in the future to combine different physical phenomena such as fluid flow, electric fields, heat transfer and surface reactions in one simulation [108]. Different new simulation approaches (e.g. for blading manufacture) are now being implemented for the first time and will be further improved.

The 2D-simulation of the DC-ECM manufacture of blades is shown in Fig. 11. The cathodes with a pre-machined geometry are moved with a constant feed rate towards the blade and due to the local conditions of electrolysis, the blade is formed. The simulation results show that the flow surfaces are closely mapped (maximum deviation <10 μm) in comparison to the experimental target geometry. Geometrical deviations of the order of 200 μm for trailing and leading edge can be explained by inaccuracies of the pre-determined inflow conductivity and inaccuracies in spline generation [109]. These deviations can be further minimised by optimisation steps. The final goal will be the implementation of an inverse simulation to predict the correct cathode shape depending on a given anode (blade) geometry [87].

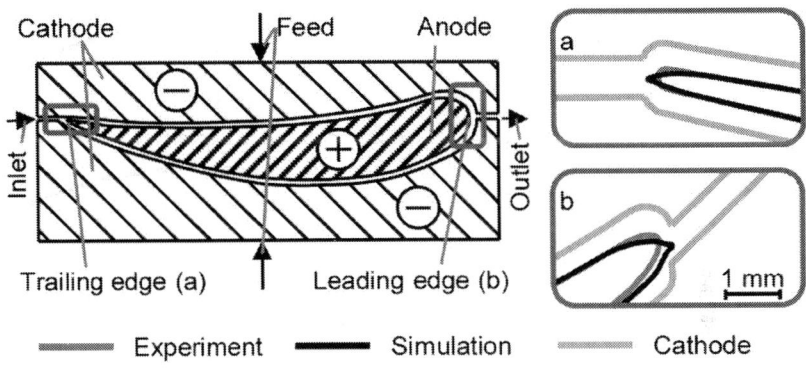

Figure.11: Principle of ECM blade manufacture and according results of a 2D multi-physics simulation. Based on [109].

In Fig. 12 a comparison between experimental results, two different 3D multi-physics simulations and the established cos( )-method (sin( ) alternatively, cf. Eq. (2) and [87]) is shown for the DC-ECM sinking of a leading edge rounded cuboid.

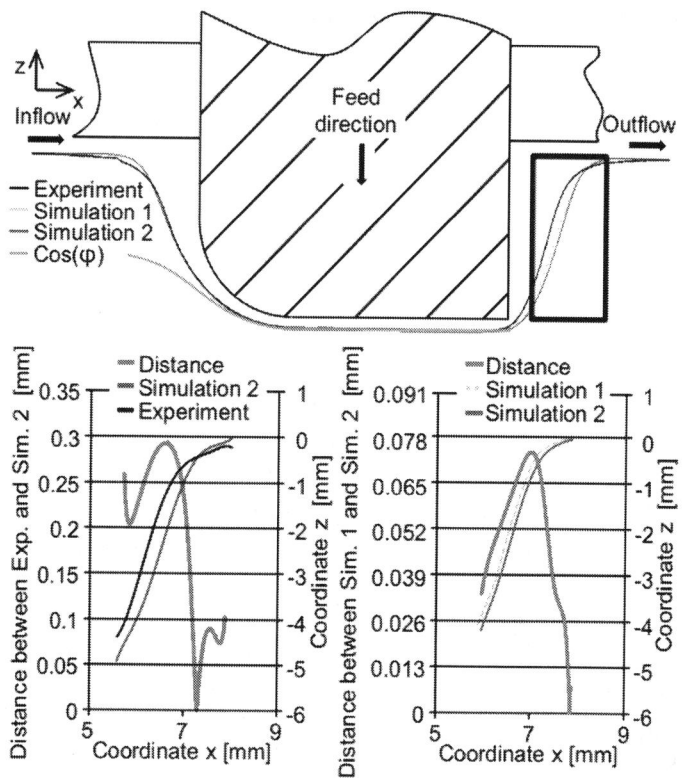

Figure.12: Modelling performance dependence on integration level of physical effects and comparison to experiment, [108].

Here the cos( )-method can predict the anode shape only for small contour angles, below 40° and therefore shows the limited capabilities of conventional ECM simulation approaches. The contour of Simulation 1 in contrast is achieved by computing the electric field in conjunction with the Faraday model. In Simulation 2 heat transfer and fluid flow are also modelled [108]. It can be seen that both simulated results show good agreement with the experiment at the inflow and the frontal gap area. Downstream, after the cathode's sharp edge, the deviations between the contours grow (shown in black box). The largest difference between the calculated and the experimental result is located in the side gap at the outflow position. The maximum deviation between the experimental and simulated results is below 300 µm. In general, the material dissolving in the simulation is slightly higher than in the experiment. This effect is caused by neglecting the hydrogen evolution, which would gradually reduce the conductivity of the electrolyte in the direction of flow. Comparing the results from both simulations, it can be stated that because of Joule heating and the linear correlation of conductivity and temperature, material removal rate is higher for Simulation 2, [108]. The approaches presented clearly show the large potential of multi-physics simulation for implementing and optimising computer-aided cathode design.

Pulsed Electrochemical Machining (PECM)

For the achievement of higher precision and better surface qualities during ECM, pulsed process modifications have been developed since about 2000. Pulsed (also termed "precise") ECM (PECM) is a vibration assisted development of ECM die-sinking by applying a low frequency oscillation of the tool electrode within the working gap [122] and [168]. Using the combination of an additionally pulsed, high current density direct current and an oscillating electrode enables precise machining at reduced working gaps of about 10 µm to 100 µm compared to typical values of 100–1000 µm during DC-ECM with surface roughnesses down to $Ra = 20$–30 nm [35]. Depending on the workpiece material, even polished surfaces can be achieved via ECM. The principle of PECM is shown in Fig. 13.

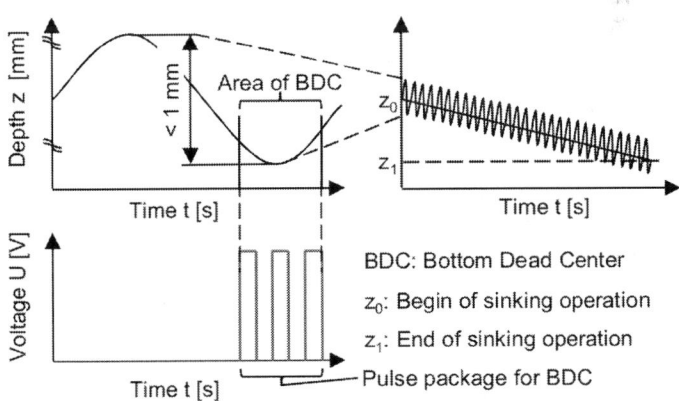

Figure.13: Principle of Pulsed (Precise) Electrochemical Machining (PECM) with oscillating cathode tool electrode.

Due to the off-times in PECM, a considerable amount of the process time is used for the replacement of the electrolyte in the pulse pauses. In these pauses, no removal process takes place. Hence, the maximum removal rates of PECM compared to EC methods, in which continuous DC current is used, are significantly reduced (e.g. 20 times longer machining times [152]). However, due to refreshment of the fluid the electrochemical conditions within the working gap are kept much more constant. Therefore, for long fluid flow paths such as during the machining of blades or other macro geometrical features of turbomachinery components, gap widening effects over the flow channels length are significantly reduced. Thus, a more uniform gap size distribution is achieved allowing more precise machining and also simplification of tool electrode development iterations. In addition, during off-times the electrolyte saturation level with ions is reduced, allowing the application of higher current densities during on-times compared to DC applications. Another advantage is the fact that due to the high current densities in the small frontal gaps, reduced stray currents occur and lower etching attack next to the machining areas takes place [152].

Specific Machine Tools and Process Handling Technology

Modern PECM machine tools allow roughing, finishing and polishing on one platform [63]. Fig. 14 shows such a machine tool with typically 7–8 axis (3–4 axis for the manipulation of the workpiece and 4 axis for tool movement (2 oscillating cathodes) [152]). The machine tool consists of a compact and closed system with autonomous electrolyte management.

Figure.14: PECM machine tool for rough and finish machining of blisk geometries with 2 vibrating cathode tool electrodes [59] and [74].

For the machining of titanium-based alloys, the electrolyte is typically aqueous solution of NaCl (avoiding passivation phenomena) while for nickel-based superalloys and TiAl, $NaNO_3$ is used. Fluid flow rates up to 1000 l/min and pressures up to 40 bar are realised in a closed-loop system with control of temperature and pH-value (via acid-/base-dosage) [59]. In addition, a monitoring and chemical treatment of chromium VI (reduction to chromium III) is nowadays implemented [195]. For filtering of the hydroxide precipitation usually back-flushable slurry filter membranes and chamber filter presses are used [59]. Finally, recycling and/or disposal have to be managed by certified specialists. Resulting costs for the electrolyte handling and the recycling/disposal of the slurry have therefore to be taken into account when considering machine investment and operation. An ecologically acceptable and user-friendly ECM process can then be realised.

The electrical power supply usually features scalable generator technology with up to 40 kA at pulse frequencies up to 10 kHz [59]. During roughing operations with constant DC, current densities of up to 3 A/mm^2 can successfully be applied allowing sufficient eduction of ions within the electrolyte [179]. Scaling with large area ECM (up to 60 cm^2[20]) achieves high material removal rates but fluid flow paths have to be kept as short as possible in order to guarantee a fast electrolyte exchange. Additionally, tool systems including fluid chambers with inlet and outlet pressures have to be designed in such a way that flow striations due to unsteady flow can be avoided. Typically, feed rates of up to 3 mm/min are possible [35].Fig. 15 shows machine set-ups for blade and blisk manufacture (DC-ECM). On the left, the simultaneous machining of 4 blades is presented. The feed rate is independent of the area to be machined enabling a significant increase in productivity [61]. On the right the blisk fluid flow box which is completely flooded during machining is shown together with the tool electrodes.

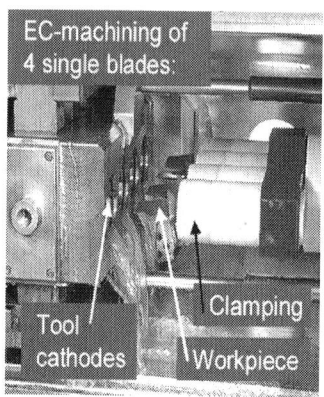 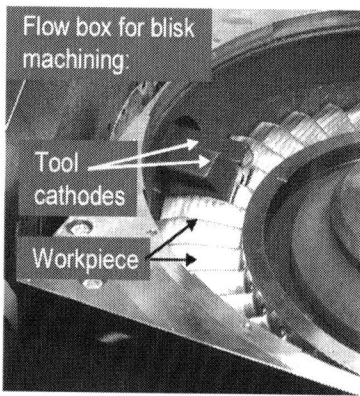

Figure.15: Machine set-up for blade (left) and blisk (right) manufacture (DC-ECM), Leistritz Turbomaschinen Technik LTT [104].

The mechanical tool oscillators of PECM have typical amplitudes of 100 mm at frequencies of up to 100 Hz. During mechanically pulsed machining, typical feed rates are only 0.1 mm/min. Therefore, the PECM process is mainly used for finish machining and it becomes economic if it is combined with preliminary steps of DC-ECM [35] and [152]. In addition to providing high material removal rates and low surface roughness, both ECM and PECM produce burr-free geometries independent of shape complexity [73].

Once a process is configured and all electrical and fluid process parameters are maintained, excellent repeatability can be achieved. During machining of a blisk with 75 blades, 43 c_p values (defined as the number of times the spread of the process fits into the tolerance band) describing different geometrical features for blade position and thickness, chord length and line angles, thickness as well as geometry of leading and trailing edges, were analysed. All values were better than 1.3 and only one was lower than 1.33 (4σ). Maximum c_p index values were around 6.65 indicating a highly stable process [152].

Examples for Successful Technological Applications

This section gives an overview of different successful application examples of ECM and PECM for the manufacture of turbomachinery components including information on relevant process conditions and achievable machining performance.

ECM is capable of producing single blade and vane geometries of different shapes both for aircraft engine and stationary steam and gas turbine applications (Fig. 16). During DC-ECM machining typical material removal can be several cm^3/min for different steel-based (e.g. X12CrNiWTiB16-13: 2.5 cm^3/min), titanium-based (e.g. Ti6242: 3.9 cm^3/min) and nickel-based (e.g. Inconel 718: 2.1 cm^3/min) alloys [20] and [72]. Form accuracy and surface roughness values of 0.1 mm and Ra = 0.8 μm respectively are possible. Calculated savings during manufacture of single blades amounts to 30% in comparison to cutting operations [35]. This is especially so in continuous production due to reduced tooling costs. Besides machining of free-form aerofoil shapes the process is also capable of machining the annulus with varying fillet radii and elliptical edges [20]. Further increase of productivity is possible during subsequent process optimisation [55]. The introduction of forged or cast TiAl blades in future engines will add even more significance to ECM [93].

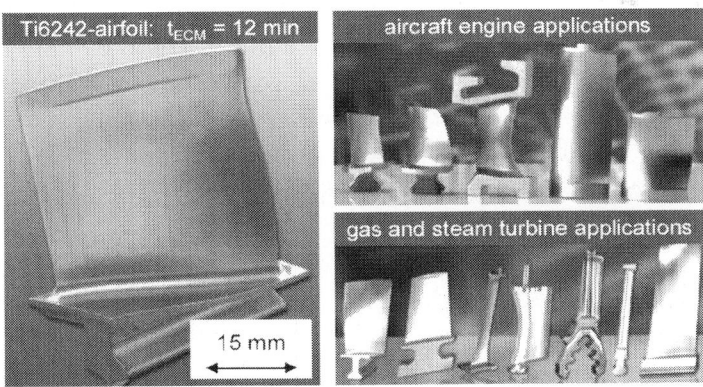

Figure. 16: EC machining of blades and vanes for aerospace and stationary gas and steam turbine applications, [20], [43] and [72].

Fig. 17 shows the machine set-up for the DC-ECM machining of blading on steel rotor shafts for stationary gas and steam turbines. The workpiece (X22CrMoV211) has a total length of 2300 mm and a diameter of 600 mm In total 120 blades are machined from 2 solid pre-turned shoulders of the shaft. Each blade is manufactured by a radial movement of a sheet electrode which is isolated underneath the tool in order to avoid extensive side gap widening. The feed rate amounts to 3 mm/min [62].

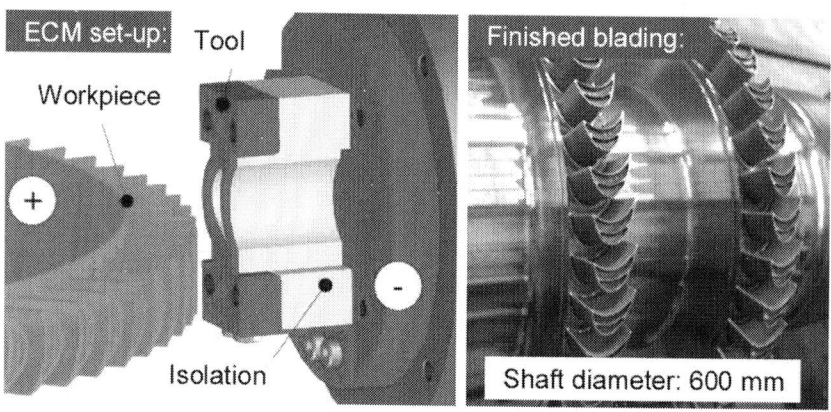

Figure.17: EC machining of blades on steel (X22CrMoV211) rotor shafts for stationary gas and steam turbine applications [60] and [62].

The EC-machining of structural components like blade flanges on jet-engine casings (diameter of 350 mm, height of 350 mm) as well as machining to facilitate weight reduction at bolt holes on disks is presented inFig. 18. In both cases, nickel-based alloys are machined via constant DC-ECM with working voltages of 20 V and 14 V and operating currents of 15 and 12 kA respectively [62].

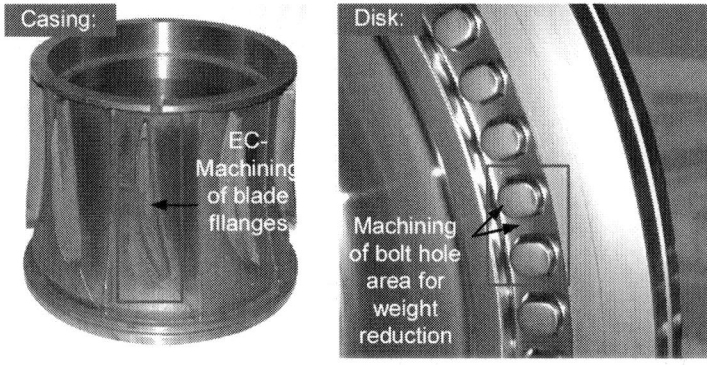

Figure. 18: EC-machining of blade flanges on jet-engine casing (left) and weight reduction at bolt holes on turbine disc (right), [62].

Fig. 19 shows an EC-machined titanium-based low pressure compressor blisk (Ti–6Al–4V) and the associated copper tool electrodes. The part has a diameter of 650 mm and incorporates 40 blades with a length of 100 mm and a profile length of 72 mm in total. It was pre-milled with an oversize of 2 mm. For the DC process NaCl was used as the electrolyte with current densities of 0.5–1 A/mm^2.

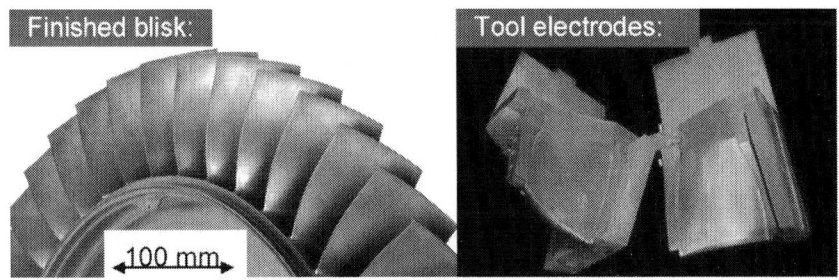

Figure. 19: Low pressure compressor blisk out of Ti6Al4V machined via ECM, [62] and [179].

Operating voltages of 13–18 V, currents of 12–15 kA and a feed rate of 1 mm/min resulted in total machining time of 5 min per blade [62], [104] and [181].

Continuous developments in cutting technology has outpaced the application of ECM for the machining of Ti-based blisks but it is still very competitive for machining of Ni-based alloys (cf. Section 4). In order to obtain the best surface qualities of aerofoils, a combination of ECM roughing and PECM finishing is suggested, the approach developed from machining a nickel-based high pressure (HP) compressor blisk made from Inconel 718 (application: preparation for series production of PW6000 HPC stage 8). The design includes thin, heavy warped blades which would be very difficult to be conventionally milled but represent ideal contours for ECM machining technologies using NaNO$_3$ as electrolyte. The machining sequence comprises DC ECM pre-machining of basic slots between the blades starting from a turned raw part. The machine set-up and the tool electrode are shown in the upper part of Fig. 20. The oversize around each blade is unequal and differs from about 1–3 mm Sufficient gap is necessary between two blades in order to accommodate the finishing electrodes. This process is followed by an unpulsed ECM roughing step to an equidistant oversize. After this, the leading and trailing edges are prepared. In the last step, PECM finishing with oscillating tool electrodes takes place to finalise the aerofoils and annulus. The machining set-up and electrodes are shown in the lower part of Fig. 20[55] and [152].

Figure.20: Machine set-up and electrode design for DC-ECM roughing and PECM finishing of blisk geometries. Based on [55] and [152].

The machining results for both ECM technologies are presented on the left and middle part of Fig. 21. In order to get a shiny, "hyper-polished" surface, an additional smoothing operation was performed with removal of oxide particles via vibratory polishing with chemical support. Final roughness values were Ra < 0.1 μm and Rz < 1 μm, see right hand photograph in Fig. 21[152].

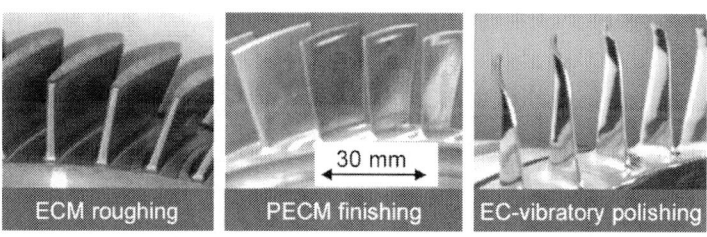

Figure 21: Production of a nickel-based HP compressor blisk via ECM, PECM and (electro-) chemical supported vibratory polishing,[36] and [152].

Besides complex blisk geometries, complete turbine wheels for turbo charger applications can also be successfully machined via ECM from solid or from near net shape, see Fig. 22[72].

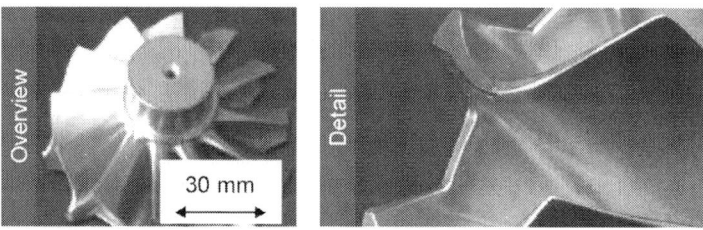

Figure 22: ECM machining of complex geometries like complete turbine wheels for turbolader charger applications [72].

EC-Machining of Cooling Holes

ECM technology is also applied in turbomachinery component manufacture for the production of cooling holes. This includes holes both for blades/vanes as well as disks. The major advantages of applying ECM are the production of smooth, stress- and crack-free surfaces.

In addition, low contour drilling angles can easily be realised [187]. Fig. 23 shows the EC machining of curved elliptical cooling holes in nickel-based high pressure turbine disks. The long axis of the ellipse has a length of 6.5 mm while the disc has a diameter of 500 mm The curved design is necessary for an ideal stress distribution during load and cannot be machined via conventional cutting. Both inlet and outlet contours are simultaneously chamfered. The overall machining time for 74 cooling holes amounts to 20 h [83] and [104].

Figure.23: Machining of curved elliptical cooling holes in nickel-based high pressure turbine disks via ECM, [60] and [83].

For machining of filigree and high aspect ratio cooling holes in blades, two ECM-based technologies are used in industry. The first process is the Shaped Tube Electrolytic Machining (STEM). The principle is shown in Fig. 24.

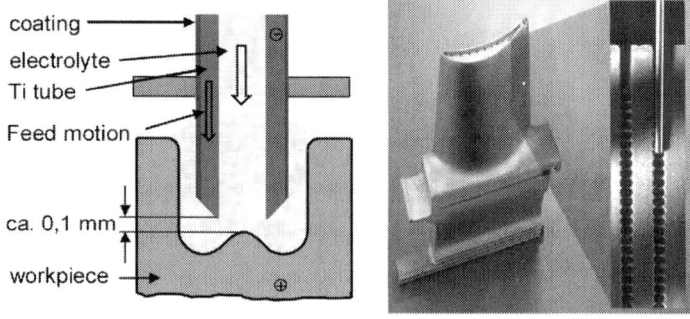

Figure.24: Principle of conventional STEM and STEM-drilling of turbulated cooling holes via feed rate variation. Based on [62] and [85].

Because of a side wall insulation of the metallic tubes used as tool electrodes, the machining only takes place in the frontal gap area avoiding side gap widening effects. High aspect ratio holes (up to 600) can therefore be drilled for diameters of 0.3–5 mm. Achievable diameter tolerances are in the range of ±0.03 mm. By using acid based electrolytes like HNO_3 and H_2SO_4, the dissolved workpiece ions are kept in the fluid allowing efficient flushing of the bore holes. Working voltages between 6 and 15 V allow feed rates of up to 4 mm/min. In addition, by periodic reduction of the feed rate, cooling holes with spherical undercuts can be produced. These so called "turbulated" cooling holes achieve a higher degree of efficiency due to the creation of swirls within the air flow [149]. Detailed experimental studies on the influence of process parameters and subsequent process modelling can be found in several publications [4], [96] and [200].

An application example of STEM drilling is shown in Fig. 25. Seven cooling holes per blade are simultaneously drilled in two workpieces within one machine set-up. The drilling length is 70 mm and the diameters range between 0.7 and 1.3 mm [62].

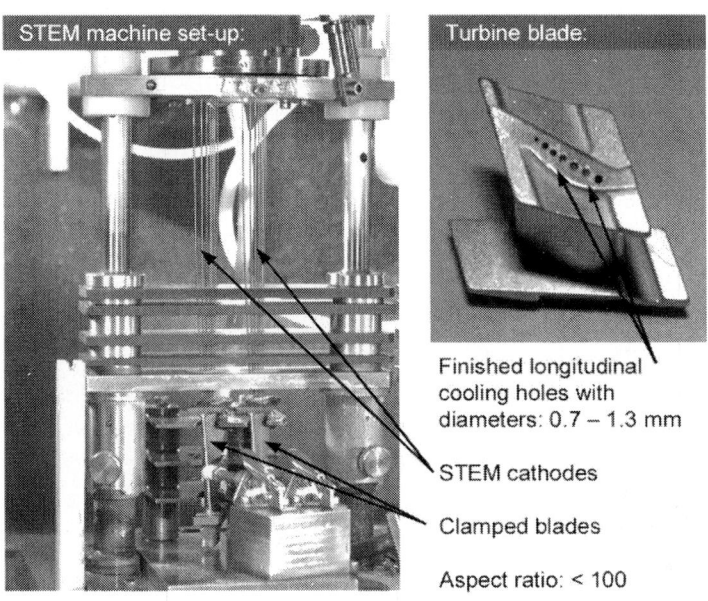

Figure.25: Machine set-up for STEM drilling of turbine rotor blades and application example [62].

The second high aspect ratio electrochemical drilling method used in aerospace industry is the Electrochemical Fine Drilling (ECF) process, [187]. Its principle and an example of application are shown inFig. 26.

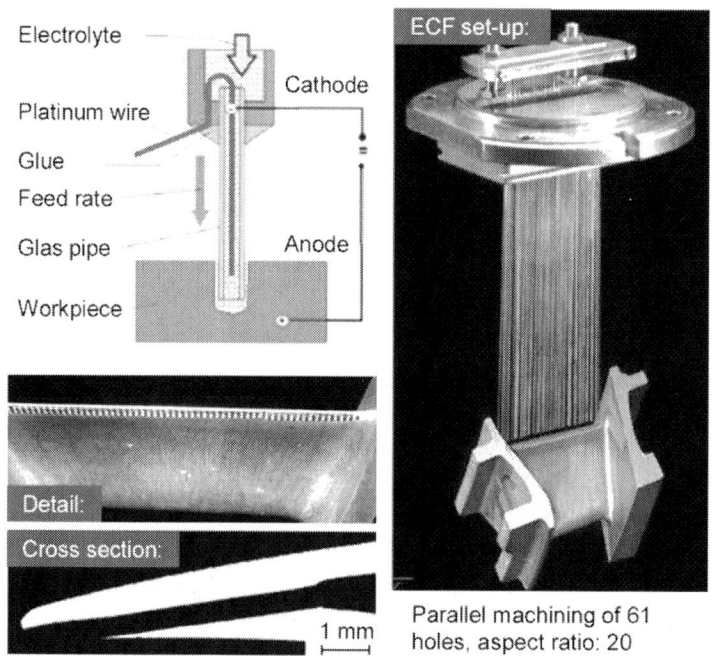

Figure.26: Electrochemical Fine Drilling (ECF) of cooling holes in nozzle guide vanes – principle and application example [62] and [187].

ECF utilises stiff glass pipes within guidance systems. Very small diameters down to 0.15 mm with tolerances in the region of ±0.01 mm for aspect ratios up to 600 can be realised. Within the pipe a metallic wire electrode serves as the cathode with working voltages of 60–100 V and feed rates of 1–4 mm/min. Electrolytes such as HNO_3 are used. The example shows the parallel machining of 61 holes with a diameter of 0.5 mm and a drilling depth of about 10 mm within a guide vane (application RB 199). The drilling angle with respect to the workpiece contour amounts to 10° [62] and [187].

ECM Grinding of Honeycomb Structures

The hybrid process combination of ECM and grinding ECG (see Lauwers et al. [122]) was developed in the 1960s in order to effect efficient burr-free material removal for difficult-to-machine aerospace alloys. The process combination allows the burr-free grinding of turbomachinery honeycomb structures or removal of heat affected zones, see Hascalik and Caydas [79]. The highly complex nature of the process and environmental concerns has led to alternative technologies being developed, limiting the application of this process.

Further Research Activities and Future Perspectives

Research activities for ECM drilling of high aspect ratio holes focus on the avoidance of acid handling by using mixed electrolytes and appropriate process parameter adaptations, see [28] and [185]. Finally, Sen and Shan [171] reported a detailed review on electrochemical macro- to micro-hole drilling processes.

The application of statistical tools (Taguchi method) for the design and optimisation of process parameters during ECM of Inconel 625 was conducted in order to define best electrolyte concentrations, feed rates and working voltages for maximum material removal rates and minimum overcut [39]. Generic aspects of tool design for ECM have been summarised by Westley et al. [208] in order to identify factors such as insulation requirements to improve machining accuracy. The design of appropriate tool electrode feeding paths and trajectory control strategies for complex workpiece geometries such as twisted blade geometries on blisks based on NC-simulation, was introduced in order to avoid interferences, irregularities of machining allowance and short circuits during exit [226] and [229]. For very thin electrodes which are necessary for narrow slot geometries (e.g. in guide vane segments), tool vibrations due to the high pressure electrolyte flow can arise resulting in poor reproducibility and high risk for short circuits [19].

Specific research on generator development has also been conducted for the electrical pulsed machining of titanium alloys in order to avoid and minimise passivation effects due to the natural oxide film formation, particularly where the change of the electrolyte system to NaCl was insufficient [204]. Similarly, related work focussing on the enhancement of pulse accuracy by MOSFET-based generator improvements has been undertaken [199]. A controlled current rise (up to 600 A/μs) with a special ramp unit allows short pulse durations <20 μs for highest currents [214].

In order to obtain higher material removal rates and best surface integrities both for ECM and PECM technologies, further comprehensive and material specific process adaptations are required. Current R&D work focusses on the detailed process analysis of different electrolyte systems and machining conditions for the machining of -TiAl intermetallic alloys with different compositions [42] and nickel-based single-crystalline materials such as LEK94, where the local microstructure and chemical composition of the material are inhomogeneous [33]. Alloying additions with high atomic number, such as W and Re, are preferentially located within the dendrites, whereas Ti and Ta are depleted. Current densities >1 A/mm^2 therefore yield more homogeneous dissolution rates [34]. A high efficiency process for pre-machining/slotting of blisk channels was developed by synchronous motion of simple tool tubes in order to finish 3 gaps simultaneously. To obtain a steady flow condition, the tubes were designed with multi-slit outlets of different width on the outer barrel. The resulting trajectories met the requirements of twisted channels and kept allowances uniform [215].

The broad capabilities of ECM/PECM, when combined in one machine set-up, to achieve both high MRR and good surface integrity together with high productivity is anticipated to see wider use in the future, especially for the machining of new advanced difficult-to-cut alloys. In this context, ECM with plate electrodes moving along the desired contour or even wire-ECM applications (e.g. machining of fir tree profiles on blades/disks) has significant potential for achieving the highest efficiencies, as reflected in its nascent use in industrial applications.

Electrical Discharge Machining (EDM)

Introduction

Familiarity with the electrical discharge machining (EDM) process belies the fact that its introduction was relatively recent, machines utilising this mode of material removal appearing only in the 1950s. For most users, the process entails either die-sink or wire EDM arrangements, which do indeed form the majority of industrial installations however there are a surprising number of alternative machine tool/process configurations that use EDM [88], [119] and [148]. Fig. 27 shows a timeline of technology developments since its inception encompassing fundamental research in addition to innovation and commercialisation.

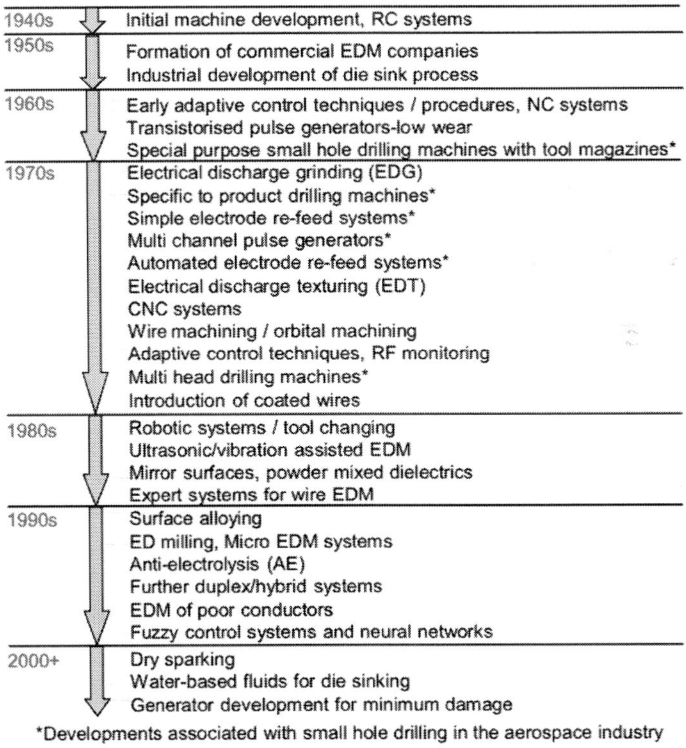

1940s	Initial machine development, RC systems
1950s	Formation of commercial EDM companies Industrial development of die sink process
1960s	Early adaptive control techniques / procedures, NC systems Transistorised pulse generators-low wear Special purpose small hole drilling machines with tool magazines*
1970s	Electrical discharge grinding (EDG) Specific to product drilling machines* Simple electrode re-feed systems* Multi channel pulse generators* Automated electrode re-feed systems* Electrical discharge texturing (EDT) CNC systems Wire machining / orbital machining Adaptive control techniques, RF monitoring Multi head drilling machines* Introduction of coated wires
1980s	Robotic systems / tool changing Ultrasonic/vibration assisted EDM Mirror surfaces, powder mixed dielectrics Expert systems for wire EDM
1990s	Surface alloying ED milling, Micro EDM systems Anti-electrolysis (AE) Further duplex/hybrid systems EDM of poor conductors Fuzzy control systems and neural networks
2000+	Dry sparking Water-based fluids for die sinking Generator development for minimum damage

*Developments associated with small hole drilling in the aerospace industry

Figure.27: Timeline of EDM process developments. Based on [8].

Aspects marked with an asterisk relate specifically to key developments associated with EDM small hole drilling in the aerospace industry [189]. Configurations such as electrical discharge grinding (in general a misnomer as the majority of commercial systems involve no abrasive action), have been adopted for superhard (PCD) tool fabrication while texturing systems are used principally for the formatting of rolls in the cold rolling of steel and aluminium sheet/strip, although recent work on texturing allied with surface alloying for fabrication of diffusion bonds in Ti–6Al–4V, has highlighted blisk or blade repair as a possible future application area [17], [124] and [144]. In contrast, dry sparking has yet to see significant application outside the laboratory environment. Since ~2000, the pace of innovation has slowed with incremental rather than step change developments, however it is likely that the introduction of minimum damage generator designs will prove a key milestone in terms of wider Wire-EDM (WEDM) application.

The largest concentration of EDM technology outside general engineering remains in the mould and die industry despite the impact of high speed end/ball end milling in the 1990s. Applications of EDM in turbomachinery manufacture whether for aircraft jet engines or industrial land based gas turbines used for power generation have not changed significantly in the past 40 years. In part, this is due to perceived adverse workpiece surface integrity issues, which are understandably a critical consideration in the aerospace industry where the main tenet and focus is on passenger safety [97]. Received wisdom [137] is such that even today, despite the developments that have taken place in generator technology (see Section 3.2.2), the thermal nature of the EDM process and the resulting workpiece damage, with consequent effects relating to fatigue life and performance, have stifled EDM expansion. Additionally, the development and take-up of laser systems for rapid hole drilling (initially partly dismissed due to shortcomings in accuracy and hole quality), have limited/restricted EDM utilisation. Notwithstanding this, EDM provides better regulation of breakthrough detection/depth control and higher achievable aspect ratios compared to laser systems, [65].

Company specific codes of practice relating to workpiece integrity acceptance standards for the different metallurgical anomalies or surface/subsurface conditions and residual stress states caused by conventional chip forming processes and non-traditional machining processes such as EDM, are operated by all turbomachinery manufacturers, however such information is proprietary and closely monitored. Predictive modelling of workpiece integrity following machining is gathering pace [97] although progress with EDM lags that of turning or milling, as does work involving aerospace alloys as compared to steels. Recent work to predict recast layer thickness/distribution has however shown good correlation with experimental observations when EDM die sinking Inconel 718 [94].

The reasons for choosing EDM over other more conventional processes are that productivity is not limited by the hardness or strength of the workpiece and complex features, or high aspect ratio holes and cavities, can be readily machined. The turbomachinery materials therefore specifically machined by EDM consist of the superalloys Inconel 738, Inconel 939, CMSX4, MAR-M002, MAR-M247, Udimet 720, Nimonic 105, Nimonic 713, etc. The main areas for application include the drilling of cooling holes and die sinking of slots, pockets and grooves (see Section 3.2.3), together with some currently limited wire cutting operations (see Section 3.2.4). Fig. 28 details sample components and ED machined features.

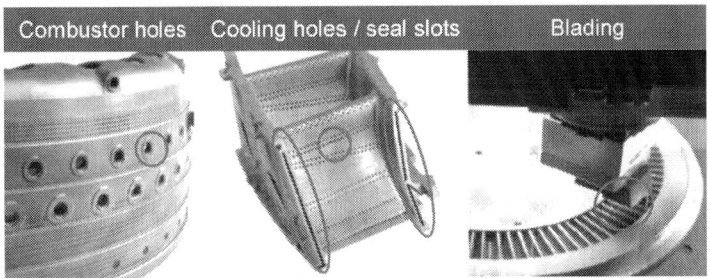

Figure.28: Sinking-EDM sample turbomachinery features indicated in red (courtesy of Winbro Group, MTU Aero Engines, GF Machine Solutions).

Hardware Considerations

The two fundamental configurations that have essentially defined the development of EDM generator technology are relaxation and transistor-based pulse generators. Relaxation generators were the first to be introduced and remained popular due to their simplicity and ability to produce both high and very low pulse energies and discharge durations, making them suitable for roughing and finishing operations as well as accurate precision machining. Conversely, transistor-based systems offered the advantage of programmable pulse shape and greater flexibility in terms of peak current, pulse width and current ignition slope settings, enabling substantial benefits in terms of increased material removal rate, reduced recast layer thickness/heat affected zone depth and application specific workpiece surface quality [9].

Historically, transistor based generators originated from the evolution of relaxation type systems towards controlled-pulse generators, where a separate AC or DC circuit is used to produce high-frequency pulses and control the relaxation-discharge energy independently [170]. For WEDM, removal rates can be increased by producing high current peaks and short pulse durations and special transistor circuits were designed for this purpose in the 1980s using high power MOSFETs. These do not use capacitors but create high current levels by controlling the direction of current supplied to the gap from high power units. For die-sinking EDM, discharge durations are longer than with wire machining which can favour lower electrode wear, but the same principle can be applied to maintain a given current level by alternating current delivery from the power source at high frequency. For finishing sequences, relaxation discharges are commonly employed, as the goal is to minimise both the peak current and duration of machining sparks. Such high frequency, low energy discharges can also be produced however by controlled-pulse circuits using only transistors, the pulse energy being defined by the gap capacitance.

Modern transistor-based generators use high power MOSFET transistors and ultra-fast recovery diodes for generating peak currents up to 1000 A with durations in the range of microseconds. In order to achieve such performance, state-of-the-art machines have very low line and machining zone inductances below 0.5 µH, allowing rising current slopes up to 600 A/µs. The transistor-based circuits can also be designed to produce trapezoidal pulses for increasing the pulse energy for roughing operations, see Fig. 29. With wire cutting systems, material removal rates on straight cuts with ferrous alloys of up to 600 mm²/min are possible when employing such configurations. However, triangular pulses with extremely short durations are often preferred in aerospace as they minimise the heat transferred to the work-piece and produce high integrity surfaces devoid of cracks and with reduced tensile stress after finishing operations. For applications involving intermediate roughness (between 0.15 and 0.8 µm Ra), similar results can be achieved with modern capacitor-based generators by exploiting the line to the machining zone as a source of capacitance and as a means for minimising the circuit inductance, thus achieving very high ratios of peak current to pulse width [51]. For polishing operations and micro-machining, very high speed MOSFET and recovery diodes are used for achieving pulse-widths of the order of 30 ns at repetition frequencies of the order of 10 MHz, with peak current values around 1 A [77]. An example of such a circuit is shown in Fig. 30. Under such conditions, surface roughness down to 0.03 µm Ra can be achieved in tungsten carbides and 0.08 µm Ra in steel, with almost zero white layer thicknesses.

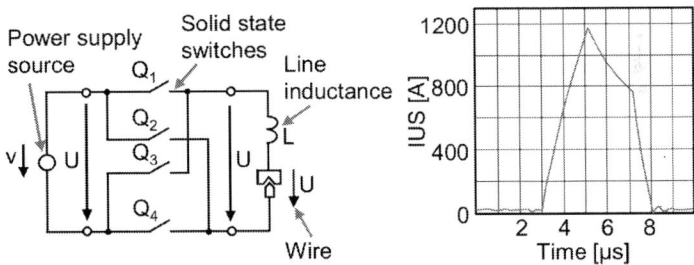

Figure.29: Basic arrangement of modern transistor-based generator and example of maximum peak current pulse for rough cutting (courtesy of GF Machine Solutions).

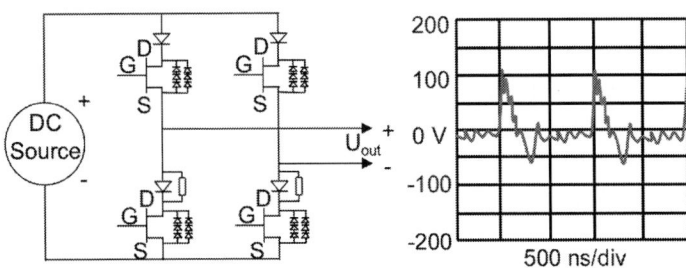

Figure.30: Principle of generic high frequency anti-electrolysis generator and corresponding sub-µs pulse profile (courtesy GF Machine Solutions).

Other key features of modern generators relate to their ability to tackle corrosion problems by introducing alternative ignition voltages such that the mean value during operation is nil (Fig. 30). Such anti-electrolysis configurations are currently used for the machining of tungsten carbides, steels and Ti-alloys. For die-sinking operations, pulse shape characteristics have not evolved as dramatically as in WEDM, progress being mainly at the process control level involving discrimination of spark quality and effect following signal analysis. Finishing operations have however, reached the same level as in WEDM as both technologies use the same type of hardware for final sequences.

The bulk of EDM machines used for gas turbine manufacture comprise drilling and die sink configurations (estimated ~90%), reflecting the component operations outlined in Section 3.2.1. Wire machines are used but on a limited basis for operations such as blank aerofoil tip machining, however future use for machining blade mounting slots in discs is understood to be under evaluation by a number of leading aerospace manufacturers (see Section 3.2.4). Published evidence for hybrid EDM use in production is slim other than for GE's announcements concerning Blue Arc™ (see Section 3.2.5), and while there appears to be growing academic and related interest in employing vibration assistance as a means to increase EDM productivity. Commercial duplex/combination systems allowing for example the option for EDM drilling and laser ablation in a single machine already exist and are intended for applications such as HP turbine blade and vane machining where laser ablation can be used for removing any thermal barrier coating prior to drilling parent material [209].

Machine tools designed for specific production operations are generally less versatile than their mainstream counterparts but are usually more productive for the particular task at hand. Limited flexibility is however afforded by multi axis CNC capability. An example is shown in Fig. 31, which details an EDM unit designed for machining holes (0.3–3.0 mm diameter, round or shaped/oval) and slots in turbine blades, vanes and segments. Variant models are able to accommodate large annular parts such as combustors and multi hole drilling operations involving up to 45 electrodes (depending on diameter and spacing) operating simultaneously. As with machinery intended for other manufacturing sectors, ancillary equipment/

systems for monitoring, probing and inspection are available. In other commercial EDM drilling systems, the combination of an individual tube electrode with a holder and positioning guide in a single assembly, avoids the need to insert the electrode during automatic tool changing, electrode rotation up to 1000 rpm providing improved flushing and removal of debris [143].

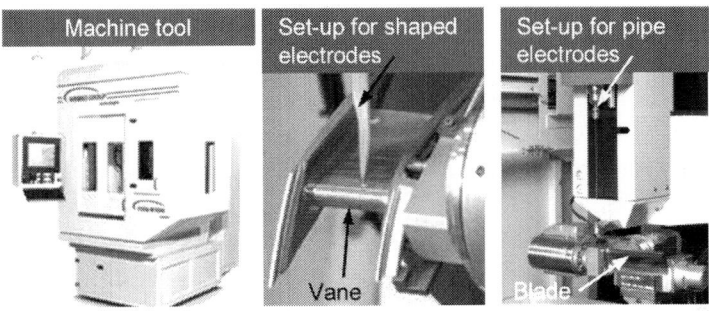

Figure.31: High speed EDM system and machine set-up for EDM drilling of blades and vanes, (courtesy of Winbro Group, GF Machine Solutions).

ED-Drilling/Die-Sinking EDM (SEDM)

The main areas for research involving die-sink EDM (SEDM) configurations in turbomachinery manufacture are twofold. The first is the machinability/optimisation of electrode type (such as graphite grade) and generator parameters when machining primarily Ti–6Al–4V [80], [102], [176], [203] and [221] as well as a number of nickel based superalloys including IN 100 [1], MAR-M247 [191], single crystal nozzle guide vane (NGV) alloy C1023 [16] and -TiAl [95]. Statistical design techniques are typically employed i.e. Taguchi, central composite designs, etc., when sinking seal slots and similar features with the output measures relating to MRR, electrode wear rate and some aspects of workpiece surface integrity (predominantly surface roughness and recast layer assessment). The second area involves the production of aerofoil blade profiles for impellers and blisks using CNC multi axis milling and electrode feeding strategies with both simple cylindrical electrodes and more complex designs. Here the work relates to development of cutter path strategies, together with bespoke machine and fixturing arrangements [56],[131] and [217]. In contrast, publications on the more fundamental aspects of EDM use with aerospace materials are minimal. An example is the work by Fonda et al. [66] concerning the influence of the thermal and electrical properties of Ti–6Al–4V on EDM productivity, where low duty factors (<10%) are shown to be optimal in order to maximise productivity and workpiece quality.

As outlined in Section 3.2.1, cooling holes, seal slots/grooves, deflector rails, damper grooves and tip recesses on shroudless blades and wedge pockets are the main features machined using EDM, with cooling holes being used extensively on turbine blades, combustors and NGV's in order to ensure adequate gas flow/cooling to enable the engine to operate at higher temperatures and hence more efficiently. Typically holes are machined in rows and groups either singularly or simultaneously with multiple electrodes, and vary in size from 0.3 to 1.0 mm, necessitating electrode diameters as small as ~0.2 mm. Corresponding wall thickness typically varies between 1 and 4 mm. Additionally, diffuser holes may be required with a 3D conical profile at hole entry. Electrode materials are predominantly solid tungsten or brass tubes with both commercial hydrocarbon oils and deionised water being used as the dielectric fluid depending on the machine system employed.

Examples of gas turbine components having features machined by sinking operations are shown in Fig. 32and include various seal slots in blades and stators as well as wedge pockets on blades. The machine tools used can be more general purpose than for drilling. Electrode materials include copper and graphite (≤10 μm grain size) with the dielectric fluid at present principally hydrocarbon oil (synthetic or mineral), although it is understood that the associated environmental issues are a concern.

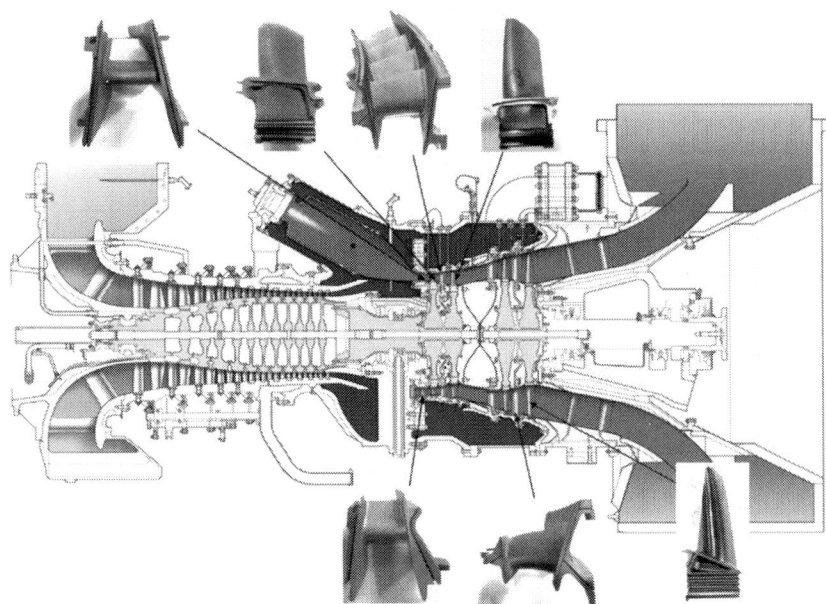

Figure.32: Industrial gas turbine components machined using EDM (courtesy of Siemens Industrial Turbomachinery).

The general capabilities of SEDM to machine dedicated turbomachinery alloys – in terms of achievable MRR – has recently been analysed at WZL in representative analogy experiments for different geometrical features (Fig. 33). Achievable MRR for constant machine set-up but material specific individually optimised machining of both blisk gaps and seal slots are presented. As basic result it can be concluded that Ni-based alloys generally possess higher maximum MRR during SEDM compared to Ti-based alloys (in contrast to conventional cutting operations). In addition, the significant increase of process capabilities in terms of achievable MRR due to latest advanced generator technology improvements are exemplarily detailed for the machining of blisk gaps for Ti-17. Comparing standard Ti-alloys and γ-TiAl it can be stated that during seal slot roughing, considerably higher MRR was achieved for 45XD compared to Ti–6Al–4V. Thus, SEDM – with a high degree of geometrical flexibility – is a viable manufacturing option for future filigree designs of components especially involving new difficult-to-cut alloys.

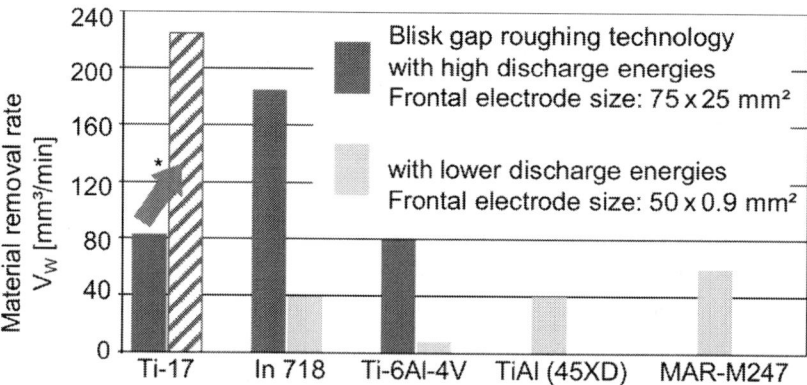

*Application of advanced generator technology compared to standard

Figure.33: Typical maximum material removal rates for Ti- and Ni-based alloys during SEDM roughing operations. Based on [113].

Wire-EDM

Developed in the late 1960s, WEDM is one of several process configurations where a continuously travelling wire is used instead of a solid tool electrode, for through section cutting of workpieces. Wires are typically uncoated or coated (predominantly zinc based for higher conductivity in the enlarged spark gap) brass, although materials such as molybdenum and tungsten are available where specific properties are required. Without doubt, the largest application area for WEDM is the mould and die industry. Other important markets include the machining of ultra-hard PCD and PCBN cutting tool blanks, biomedical instruments and precision/micro devices.

Unlike conventional ED-die sink/drilling, the use of WEDM for the manufacture of turbomachinery components is currently minimal and confined to either noncritical features or those subject to post processing operations. Established applications essentially include the machining of titanium blisk aerofoil tips, the breakthrough of nickel based superalloy stator vane rings and little else. The reluctance in adopting wider utilisation of WEDM has partly stemmed from slower material removal rates (compared to traditional cutting processes), but more crucially the prevailing perception of poor workpiece surface integrity due to the thermal nature of the process.

Aided by the significant innovations and progress in generator, wire, system control and monitoring capabilities over the last ~15 years, the case for greater uptake of WEDM by the turbomachinery sector is growing, in part prompted by the increased pace of academic/industrial research over the past decade involving nickel based superalloys and titanium alloys. Much of this early work centred largely on the use of statistically designed experiments (Taguchi, response surface methodology) and modelling/optimisation techniques (artificial neural networks, Pareto analysis) to determine preferred operating parameter levels for maximising MRR and improving workpiece surface roughness [84], [157] and [166]. The reported trends with regard to the influence of varying machining conditions on response measures were essentially similar irrespective of the workpiece material assessed. The lack of associated workpiece surface/subsurface integrity results however, prevented any meaningful assessment of process feasibility for turbomachinery components, with experiments generally confined to standard single pass cutting and the resulting surface roughness values >2 µm Ra, well above acceptable tolerances (Ra ≤ 0.8 µm) [106].

More recently, research has been carried out to evaluate the potential of WEDM for the manufacture of fir tree or dovetail profiled blade root mounting slots in turbine/compressor discs, see Fig. 34, which are presently finish machined almost exclusively by broaching. While there are relatively few alternatives for finishing slots, several viable options exist when roughing, some of which are currently used in production including conventional milling, creep feed grinding (CFG) and abrasive waterjet cutting (AWJC)[48] and [49]. When finishing, point grinding using small diameter profiled electroplated superabrasive wheels has been extensively evaluated, however the scale of industrial implementation is unknown [15]. The concept of utilising WEDM for producing fir tree root slots was briefly outlined in a paper published in the mid-1980s [47], however there were no indications that the technology was sufficiently mature at the time to be implemented commercially by any of the major aeroengine manufacturers.

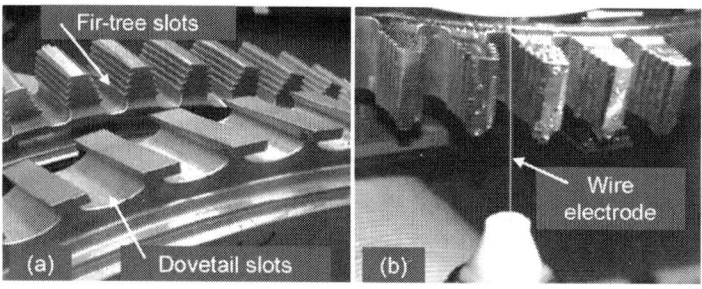

Figure.34: (a) Sample blade mounting slots profiles in discs and (b) WEDM of fir tree root slot demonstrator [177].

Aspinwall et al. [14] reported preliminary experimental work using a high specification 'minimum damage' WEDM machine to investigate the influence of a combined roughing and multiple trim/finishing cut strategy on workpiece surface integrity when machining 10 mm thick Inconel 718 and Ti–6Al–4V material. Discontinuous/non uniform recast layers were observed following the main roughing pass, with average thicknesses in the region of ~6–10 µm for Inconel 718 and Ti–6Al–4V respectively. These were found to decrease steadily with successive trim cuts and near damage free surfaces (essentially zero recast) were produced after 4 finishing passes with workpiece surface roughness between ~0.2 and 0.4 µm Ra achieved after the final trim cut. Analysis of cross sectional micrographs revealed no obvious signs of subsurface microstructural alterations or thermal degradation with minimal variation in microhardness, the low energy high frequency/short duration discharges (>1 MHz for the third and fourth finishing cuts) essentially only removing the damage from preceding cuts without inflicting any additional thermal degradation on the workpiece.

The results stimulated further in-depth research encompassing WEDM of Udimet 720 and Ti–6Al–2Sn–4Zr–6Mo (Ti-6246) aeroengine alloys [9], [10], [11] and [177]. Here workpiece thickness was 30–37 mm, which is representative of typical dimensions in disc root slots. Surface roughness and recast layer thickness following the 4 trim cut strategy did not exceed ~0.5 µm Sa and 2 µm respectively in any of the samples evaluated, the latter being easily removed using a post-etch operation. Residual stress measurements revealed a near neutral state (<90 MPa) at the machined surface for both the Udimet 720 and Ti-6246 specimens following the fourth finishing pass, despite the former being highly tensile (~500 MPa) after roughing. Furthermore, depth profile analysis showed that the tensile residual stresses did not extend beyond ~40–50 μm below the machined surface.

Following modification of generator pulse profiles and manipulation of operating parameters, tests were performed involving a rough and 2 trim cut strategy using both uncoated and zinc coated brass wires in an attempt to improve process productivity [10]. Comparable results were obtained with the updated 2 trim cut procedure, where average recast layer thickness reduced from ~8 μm after roughing to <2 μm in both the Udimet 720 and Ti-6246 workpieces, see Fig. 35. Similarly, surface roughness was maintained at ~0.6 μm Ra while surface residual stress levels were almost neutral (<80 MPa) after the second finishing pass. In addition, significant increases in removal rates were recorded (40% for Udimet 720 and 70% for Ti-6246) when roughing using coated wires over uncoated brass, with no appreciable change to workpiece surface integrity.

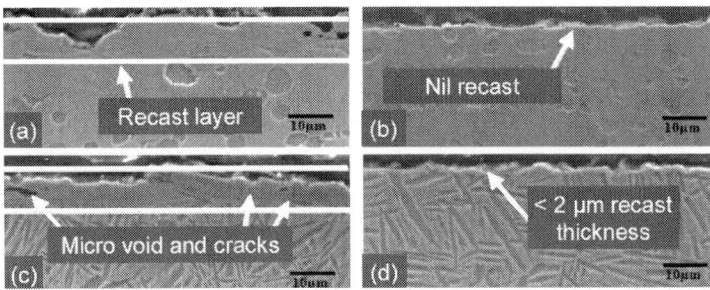

Figure.35: Cross sectional micrographs for (a) Udimet 720 Rough cut, (b) Udimet 720 Trim Cut 2, (c) Ti6246 Rough Cut, (d) Ti6246 Trim Cut 2[10].

As detailed by Jawahir et al. [97], the influence of machining operations on surface integrity and the subsequent knock on effects on fatigue strength/endurance is of paramount importance, especially where safety critical components or those subject to cyclic loading are concerned. In room temperature high cycle fatigue (HCF) tests on rough and finish (using 4 trim cut strategy) wire machined Udimet 720 specimens together with flank milled samples [11], the run-out stress of the rough WEDM specimens after 1.2×10^7 cycles was only 0.39 of the materials ultimate tensile strength (UTS), however the finish specimens recorded a value of 0.59 UTS and the flank milled samples

0.65 UTS, see Fig. 36. The marginal difference in fatigue life between the finish WEDM and milled samples was mainly attributed to the presence of compressive surface residual stress (approx. −220 MPa) in the latter, although the difference was not statistically significant. Complementary trials to assess the effect of finish WEDM on fatigue performance of Ti-6246 samples yielded similar results, see Fig. 36, [177].

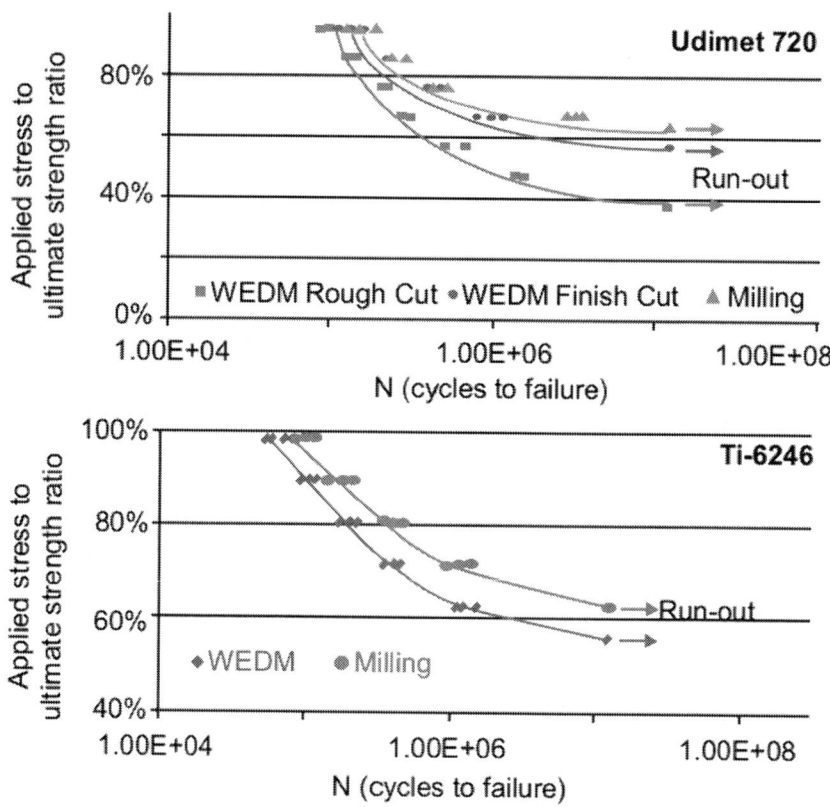

Figure.36: S–N fatigue curves for Udimet 720 and Ti-6246 specimens. Based on [11] and [177].

Parallel investigations carried out by researchers in Germany [105] to compare the fatigue life of WEDM against ground Ti–6Al–4V surfaces highlighted superior performance in the former. This was somewhat surprising, particularly as associated surface integrity assessment indicated lower surface roughness levels (0.17 vs. 0.25 μm Ra) and a smaller region of heat affected zone (8 vs. 20 μm) in the ground samples. However, high stress concentrations at the edge of sharp cracks observed on the ground surfaces was suggested as the likely reason for the poorer fatigue limit recorded.

In further work by Klocke et al. [106], the reliability and repeatability of a 3 pass WEDM strategy (rough, finish and polish cuts) developed for Inconel 718 was investigated. A series of four fir tree shaped root slots were machined sequentially using uncoated brass wire on a 40 mm thick test block, to mimic a typical turbine/compressor disc section. Cross sectional evaluation of the final cut surfaces (<0.8 μm Ra), showed no evidence of recast layer formation, cracking, porosity, phase transition or microstructural alterations. Precision checks to ascertain accuracy of the machined profile was also performed, with results showing that the cut geometry was within a tolerance band of ±5 μm, see Fig. 37. The mean cutting time per slot however was quoted at ~71 min, which is a relatively slow rate of production compared with alternative processes [113]. Process optimisation (including use of coated wires and modification of generator settings) reduced this value to 40 min, with anticipated potential for further improvement [107]. Associated high cycle fatigue bending (HCFB) tests were carried out by Welling [207] to compare the performance of the optimised WEDM technology against broaching and grinding when machining Inconel 718 samples. Despite higher mean surface roughness values on the WEDM (0.7 μm Ra) compared to broached (0.28 μm Ra) samples, the failure bending moment was found to be equivalent, see Fig. 37. The fatigue limit of both processes was however ~35% lower with respect to grinding. Not all recent work however has shown damage free surfaces following WEDM [128]. Recast layers ~3.3 μm thick with a roughness Ra of 1.25 μm after 3 trim passes are reported when machining Inconel 718. Other recent research work focusses on the analysis and optimisation of WEDM for Ti–6Al–4V and -TiAl [120] and [167].

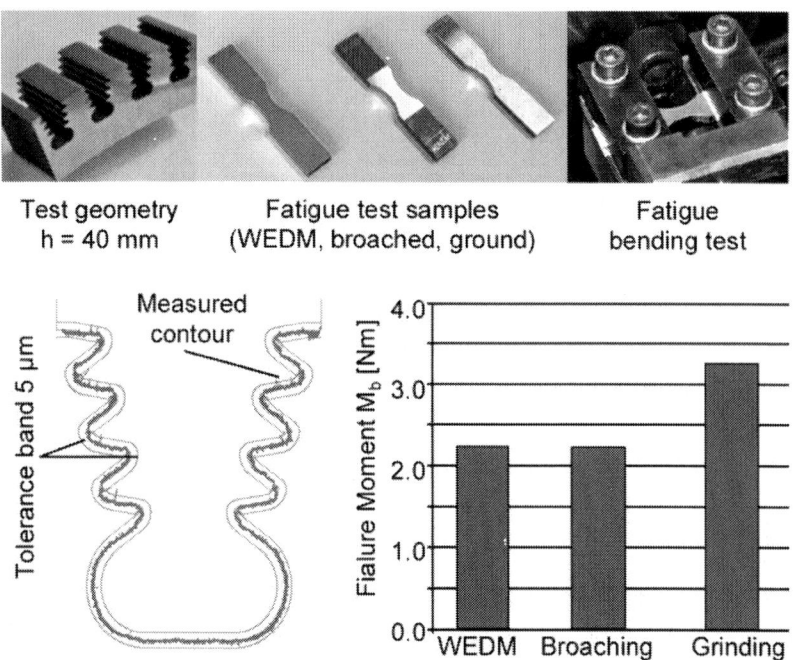

Figure.37: Precision WEDM machining of fir tree slots and analysis of workpiece functionality via bending fatigue tests, [106] and [207].

Process and Machine Tool Adaptations for Specific Tasks

The limitations of EDM in terms of material removal rate (MRR) and the dichotomy between attempts at greatly increasing productivity while maintaining or achieving an acceptable workpiece roughness and integrity, are outlined by Wei and his co-authors [205]. These include the fact that with standard EDM systems, pulse mode operation is such that material removal does not occur 100% of the time. Furthermore, an upper boundary exists in relation to the maximum current level that can be usefully applied without compromising process stability, although generator developments over the past decade have enabled ED wire machines to operate with peak currents of 1000 A and slopes of up to 600 A/μs as reported in Section 3.2.2. Even so, with die sink and related drilling operations, excessively high current levels can adversely affect workpiece topography and electrode wear, the associated crater debris producing arcing in the typically narrow inter-electrode gap, and servo instability. Here removal performance is highly dependent on electrode type and geometry, cavity shape and dielectric flushing, and while a stock removal rate of ~1 cm^3/min is possible in steel, for industrial applications involving aerospace materials such as Inconel alloys, removal rates of up to 0.2 cm^3/min (with relative wear below 0.3%) are more realistic.

Improved MRR has been reported with a number of hybrid EDM processes, the most significant relating to erosion systems employing continuous arcing or a combination of controlled arcing and discharges, as a consequence of the higher energy densities that are possible compared with spark discharges alone. Textbook entries relating to electroerosion arcing (as opposed to plasma arcing using a gas assist) as a means to remove material are limited, but where they do occur [137], arc sawing (AS) or less accurately, electrical discharge sawing (EDS), is often detailed involving a thin steel band or rotating disc electrode. However whereas standard EDM uses a dielectric fluid, AS systems generally employ a solution of highly conductive sodium silicate (water glass ~20,000 μS/cm) or alternatively water (~1000 μS/cm) or air, with a DC generator providing a low voltage (6–60 V), high current (200–15000 A) supply. Additionally the feed may be constant/uniform or servo controlled. Workpiece accuracy and roughness are less controllable than for EDM and machined surfaces are subject to greater thermal degradation which can encompass re-deposited material, cracking and a recast layer (typically 50–100 μm but can be greater depending on arcing energy and electrolyte/fluid used), with consequent changes to workpiece microstructure and microhardness.

Commercial machines utilising arcing to rapidly section large forging blanks, honeycomb structures, castings, extrusion products, bar stock, etc., in a range of difficult to cut materials including stainless steels, nickel and titanium alloys, appeared during the 1960s both in the Soviet Union and USA and subsequently in Japan [150]. In contrast to EDM development and growth however, their use was not mainstream or widespread, the main focus of application during the late 1980s and 90s appearing to be in the decommissioning of atomic reactor pressure vessels, where any workpiece damage (typically recast layer material and cracking) and relative inaccuracy resulting from the thermal erosion process, was of little concern.

Reported in several papers on high efficiency electrical erosion, the electro-melting-explosion process outlined schematically by Ye [219], bears some similarity with the AS systems reported above, having voltage and current values when rough processing of 25–28 V and 1600–3000 A respectively, but with the option of fine processing at 1–9 V and 1–90 A. The intended workpieces include nickel and titanium alloys with the preferred working fluid, a low concentration aqueous electrolyte. Gap details during non-contact processing are quoted as 0.01–0.1 mm but with acknowledgement that short duration mechanical contact between the workpiece and rotating tool disc (5–2000 mm in diameter and from 1 to 40 mm thick) may aid processing. Whether the process has been commercialised is unknown. Although limited by modest generator power availability and flushing constraints, AS work reported by Paul et al. [151] was aimed at effecting fast bulk metal removal of nickel based alloy for aeroengine component applications. In contrast to the more familiar cut off operations, the paper details profiling and slotting using 10 mm thick plain and shaped disc electrodes. Workpiece surface integrity evaluation indicated a maximum white layer thickness of 55 μm with cracks propagating to ≤40 μm.

More recent publications cite applications directly relevant to turbomachinery manufacture, in particular the rapid rough machining of aerofoil blades in blisks using high speed electroerosion machining (HSEM/Blue Arc™). Cylindrical electrodes are used with through flushing capability, material removal occurring via transient arcing through an electrolyte, though not one containing sodium silicate, see schematic in Fig. 38. Multi axis tool path movement incorporating dynamic tool wear compensation is employed to 'end mill' the aerofoils with a ball nose tool on a layer by layer basis. The figure shows an example of HSEM blade 'milling' together with sample microstructural damage on Inconel 718. The process is detailed as 15–20 times faster than EDM milling, up to 3 times faster than conventional milling and able to remove upwards of 200 cm^3/min when operated in a pseudo grinding mode with a 25 mm thick rotating disc electrode[205] and [222].

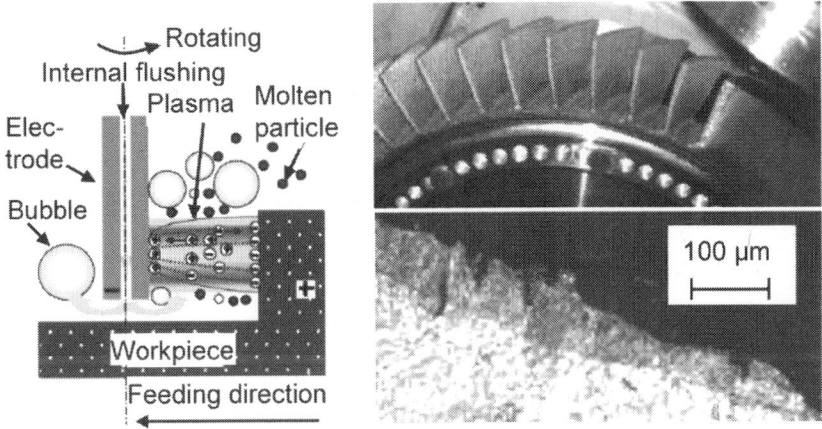

Figure. 38: Schematic of HSEM/Blue Arc™ process mechanism and machining of superalloy blisk (Inconel 718) workpiece [205] and [222].

The research reported by Zhao et al. [227] on blasting erosion arc machining (BEAM) bears some comparison through its use of either a bundle of graphite tube electrodes (Fig. 39), which can provide a 3D shaped end face or a laminated electrode with slotted surface and multiple inner holes. In cavity machining of Inconel 718, a material removal rate of 11.3 cm³/min is quoted for a current of 600 A and an on-time of 2000 μs using transitory/intermittent rather than steady state arcing. A prerequisite for successful operation is the use of high velocity flushing with the water-based dielectric to induce strong hydrodynamic forces into the gap in order to periodically distort and break the plasma column and blast away molten material. Tool wear ratio is detailed as less than 3% with the thickness of the heat affected zone less than 200 μm.

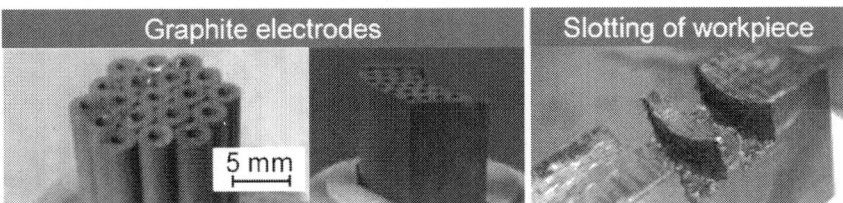

Figure. 39: Blasting erosion arc machining (BEAM) using bundled graphite electrodes for rough machining of blisk slots. Based on [227].

For increasing productivity the simultaneous machining of seal slots in different parts (fairings, panels, etc.) in one machine set-up is state-of-the-art. In addition units for vibration-assisted machining of such slots are under development (Fig. 40).

Figure.40: Simultaneous machining of several seal slots in one set-up and schematic of twin piezo graphite electrode unit for vibration assisted EDM machining with standard interface. Based on [138] and [192].

Research involving ultrasonic assisted EDM (USM/EDM) and latterly that relating to lower frequency vibration assisted EDM, has shown there to be significant benefits in drilling/die-sink operations in terms of increased MRR/reduced machining time, increased penetration depth, a reduction in the incidence of arcing together with improvement in sparking efficiency, thinner recast layer and heat affected zone and in some cases reduced tool electrode wear, principally as a result of improved flushing and clearance of debris from the spark gap. In the early 1980s, Kremer et al. [116] in trying to resolve some of the limiting issues of EDM such as poor dielectric circulation and the evacuation of debris and gasses especially with intricate electrodes, showed that using an electrode vibrating at ultrasonic frequency ~ 20 kHz allowed deeper penetrations and higher feed rates in die sink operations. Feed penetration when using a graphite electrode to machine a steel workpiece increased some 30% when roughing and 300% when finishing. Subsequent publications [117], [129], [142] and [186] all involving vibration at ultrasonic frequencies (20–23 kHz) with amplitudes ranging from 3 to 30 µm, both with and without abrasive particles dispersed in the dielectric fluid, detail positive benefits for the USM/EDM approach. The paper by Lin et al. [129] is particularly relevant, the MRR when USM/EDM Ti–6Al–4V with a dispersal of 3 µm

SiC in distilled water dielectric (90 g/l) being twice that of conventional EDM. Here the concentration of abrasive is reported as critical, with too much producing unstable discharges. Complementary research with wire machine configurations also detail improvements in performance, Guo et al. [76] reporting cutting efficiency increased by 30% and reductions in both workpiece surface roughness and tensile stress when vibrating the wire at 35 kHz with amplitudes up to 12 µm. Somewhat less striking are the results reported by Han et al. [78] when ultrasonically vibrating a Ti–6Al–4V workpiece (in cutting direction). Here the benefit in MRR compared with standard EDM is quoted as ~10%.

Results for lower frequency longitudinal tool vibration reported by Prihandana et el. [155] and Uhlmann and Domingos [192] are similarly positive, with data for the former when operating at 600 Hz and 0.75 µm amplitude with a 12.5 mm Cu tool electrode, suggesting a 23% increases in MRR and lower workpiece roughness and tool wear rate when machining SS304 stainless steel. In the latter paper, twin piezo-actuators are detailed for the machining of high aspect ratio seal slots in MAR–M247 using graphite electrodes vibrating at up to 1000 Hz with amplitudes from 2 to 16 µm achieving increased MRR and reduced wear.

Further specialised machine equipment has been developed for the flexible machining of radial stator ring cut-outs for blading fixturing with workpiece diameters up to 2000 mm via WEDM (standard precision adequate) (Fig. 41).

 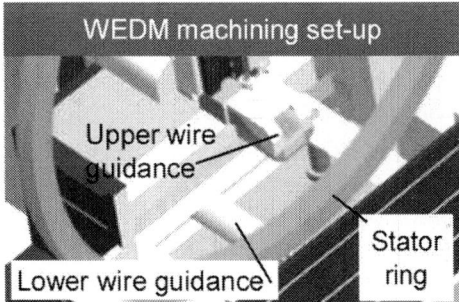

Figure 41: Machine equipment for flexible machining of blading cut-outs in stator rings, (courtesy of Exetec/Ona Electroerosion).

Outlook and Future Perspectives

With productivity as a key driver, the move to develop specialised EDM equipment in addition to adapting and optimising standard systems for the manufacture of turbo-machinery components [165] is likely to continue. There is little evidence at present of operating modes such as ED-milling, which is able to utilise simple electrodes, being used in practice, however this may change due to recent developments in hybrid processing, especially where the operation involves only roughing. The considerable volume of data generated in the past 6 years relating to surface integrity and fatigue life, clearly points to WEDM as a potentially viable process to more traditional manufacturing techniques for turbomachinery components. Additional benefits relate to the possibility of 24 h unmanned operation, the flexibility afforded by the software driven process nature in respect of part design changes together with real time condition monitoring procedures. In terms of product miniaturisation, EDM can offer advantages especially for the machining of ceramic components [130] and [228]. While minimum damage cutting of various nickel and titanium alloys has been demonstrated, producing acceptable surfaces in materials such as -TiAl using EDM is more problematic. Poor surface integrity characterised by long/deep cracks which propagate along the grain boundaries into the bulk material as described by Mantle et al. [133] in the late 1990s, remains true, even when using more up to date generator technology with multiple trim cuts.

Despite some six decades of R&D work, current use of controlled electroerosion arcing for turbomachinery manufacture appears extremely limited. Recent work has shown the possibility of significant productivity benefits in specific applications however the lack of process capital/operating cost data, the scarcity of suitable commercial machine tools and the adverse accuracy and workpiece integrity associated with the process need to be addressed for the process to have greater impact.

Photonic Processes I: Additive Manufacturing

Overview

Additive Manufacturing technologies (AM) enable the build-up of complex objects by the repeated application of thin layers, see Fig. 42. Compared to conventional subtractive processes, economic efficiency is possible with quantities starting from 1, being substantially less dependent on the part geometry. AM originated from the manufacture of prototypes (Rapid Prototyping), but is being increasingly employed for small-scale series production (Rapid Manufacturing) and tooling applications (Rapid Tooling). Deposition procedures are capable of locally applying and solidifying the base material (metallic powder or wire) on a part, rendering this technology extremely efficient for repair applications. In powder bed metal-based procedures, flat layers of base material (i.e. metallic powder) are applied for the complete build area. In a second step the current part cross section is solidified by selected heat input through laser beam melting (commonly called selective laser melting - SLM) or electron beam melting (EBM). The principal limitations of the approach include the requirement for additional support structures to ensure proper heat dissipation and to reduce distortions. Build-up of overhanging structures therefore requires advanced process knowledge and is followed by additional post treatment steps. The basic process mechanism is detailed in a paper by Kruth et al. [118]. Development of AM for aeroengine and turbomachinery production started in the 1990s [212]. Besides prototyping, tooling and jig applications, the direct manufacturing of serial parts is of major interest due to increased deposition rates as well as improved material properties [210]. Volume build rates of up to 10 cm^3/min for Inconel 718 are feasible in layer-wise laser cladding AM [69] and [211]. To overcome admission restrictions especially in the aerospace sector, a great deal of effort has been put into qualification [223], testing and certification of additively manufactured parts including development of online quality assurance systems [45].

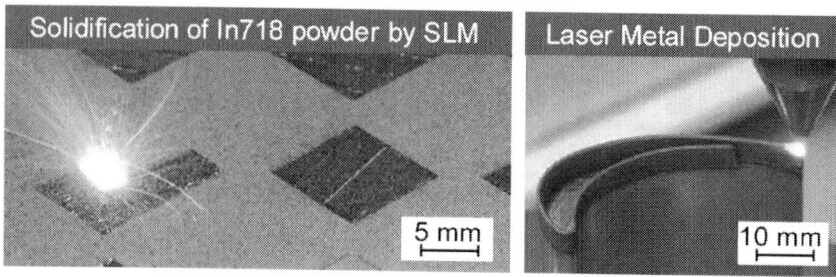

Figure 42: Additive Manufacturing by Selective Laser Melting SLM [courtesy of IWB] and Laser Metal Deposition LMD [69].

The additive manufacturing technologies, see turbomachinery examples in Fig. 43, offer new possibilities for product design (complex shapes, hollow spaces, undercuts) while simultaneously allowing for a significant cost and weight reduction [160]. MTU Aero Engines is using AM for a wide range of parts, e.g. vane segments and honeycomb structures for rigs and "ground only" engines [82]. The production of the first flying component, a boroscope eye for the PW1133G NEO started in 2013 [121].

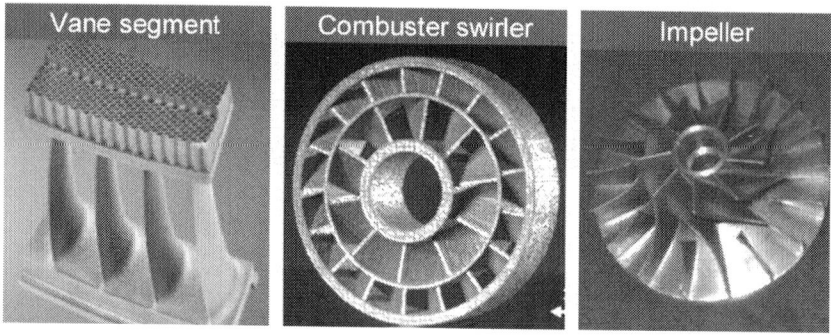

Figure.43: AM manufactured vane segment with honeycomb seal structure [54], functional prototype of In 718 combuster swirler [103] and In 718 impeller blading build on pre-machined part [216].

General Electric (GE) Aviation is also moving towards serial production and recently announced the manufacture of a leading edge for blades and a fuel injector in 2013 [213]. By 2020, GE is scheduled to produce more than 100,000 additively manu-factured components for its LEAP and GE9x engines. Of major interest is a design optimised fuel nozzle, which is ~25% lighter and five times more durable than conventionally manufactured fuel nozzles [71]. NASA uses additive technologies to build various parts of space launch systems and considers Selective Laser Melting (SLM) to be the manufacturing future [92].

Future Applications

AM is expected to have considerable potential due to increased design opportunities that include graded workpiece properties (see Fig. 44), organic shapes or geodesic structures during the manufacture of diverse turbomachinery components (blades, vanes, shrouds and according segments, blisks, casings, tubes and brackets, nozzles, liners, etc.) [3] and [6].

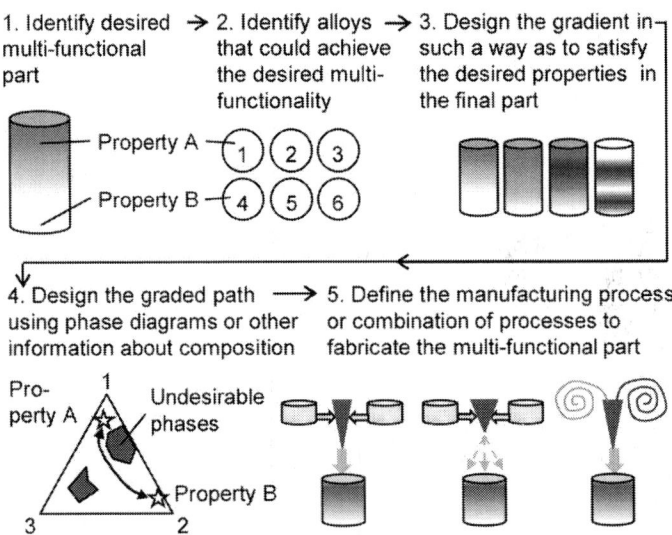

Figure.44: Method for fabricating gradient alloy parts with multi-functional properties. Based on [230].

As an example for future application, a new approach to compressor casing design was discussed [12]. Here, auxetic structures (i.e. materials with negative Poisson's ratio) are used to reduce the relative tip clearance by adapting the behaviour of the casing to that of the rotor, see Fig. 45. The reduction of relative tip clearance resulted in an increase in compressor efficiency. Due to the fact that the fabrication of auxetic structures is difficult or even impossible with conventional manufacturing techniques, the application of additive manufacturing technologies (SLM or EBM) is proposed. The design freedom provided by additive manufacturing can in the future be used for the fabrication of hollow rotor blades resulting in a decrease of total engine weight. Furthermore, the variation of wall thicknesses makes it possible to shift certain blade eigenmodes in order to reduce the total vibrational load.

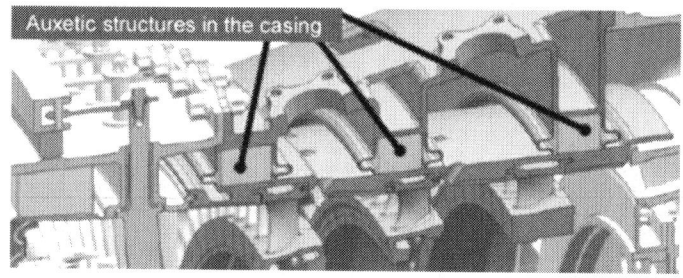

Figure.45: Position of the auxetic structures in a double-walled split casing for reducing the relative tip clearance. Based on [12].

Process Development

Additive processing of Ni- and Ti-based alloys used in turbomachinery applications is mainly done via SLM or EBM. In laser processing, the absorption of a high intensity near infrared (NIR) beam causes the metallic powder particles to melt and form a continuous bond between subsequent scan tracks while cooling down from melt temperature. Absorption is dominated by multiple reflections at individual powder particles, which makes the absorption of NIR laser beams in powder layers significantly more efficient than for bulk material. In SLM, an inert gas flow at slightly over pressure is used to prevent oxidation and to remove sputter particles ejected from the process zone. The solidification mechanisms [118] are controlled by the melt flow behaviour and wettability and can be varied by the applied scanning strategy. Since focal diameter, powder grain size and layer thickness are of the same order of magnitude, filigree turbomachinery components can be fabricated to precision of 100 µm with commercially available system technology.

During solidification the size of the melt pool is mainly determined by the heat conduction through the solid because the thermal conductivity of the surrounding metallic powder is substantially lower [5]. In case of insufficient contact between adjacent layers, process instabilities may develop causing the ejection of melt droplets and sputter particles [135]. The melt›s temperature may rise significantly above the evaporation limit during material exposure and a plasma plume is formed which exerts a recoil pressure on the melt pool and increases the sputter activity [139].

Compared to a clear focus on high temperature Ni-base, Au- and steel materials, EBM technology developed by ARCAM AB has centred mainly on titanium alloys. Zäh and Lutzmann [224] modelled the EBM process and developed understanding of the "powder blowing" effect caused by electric charging of the powder particles. The technique requires a low-pressure vacuum or inert gas atmosphere (10–1 Pa) in the processing chamber to reduce electrical charging of the surrounding powder bed [75]. By using preheating strategies, it is possible to process materials susceptible to cracking. Preheating temperatures of up to 800°C [173] are possible, enabling the reduction of support structures through reduced thermal gradients. Additionally, using a split electron beam (the so-called multi-beam strategy) allows for high speed processing compared to conventional hatching strategies used in SLM. The current system technology development in both cases focuses on high power beam sources and multiple scanning units to enable the processing of materials with low absorbance and high thermal conductivity (i.e. Cu). Advanced exposure strategies subdivide the component into skin and core areas in order to use different process parameters. The skin section is usually solidified with moderate energy input and small focal diameters to account for filigree structures and optimised surface properties, whereas the core is exposed to high energy input and big focal diameters to increase the manufacturing speed. High power SLM with tailored process control (skin-core) allows Inconel 718 components with relatively low thermal conductivity (≤ 30 W/mK) to be manufactured with a density ≥99.5%. Hence, the use of lasers with output power of up to 1 kW enables build-up rates to be increased by a factor of 4, compared to conventional SLM [136]. Advanced process and cooling strategies have also been developed for the machining of complex blisk parts [211], with typical examples shown in Fig. 46.

Figure.46: IN718 Nozzle Guide Vane studied for application of High-Power SLM [Courtesy of TurboMecca/136] and blisk demonstrator used for AM machining and cooling strategy development [211].

In many cases, EBM and SLM technologies do not fulfil requirements relating to surface quality or dimensional accuracy [44], and post processing by means of conventional machining or sandblasting becomes necessary (see Section 3.3.7). Current research activities focus on the optimisation of exposure parameters and on statistical confirmation of material properties. It was shown that comparable part quality can be obtained when moving towards higher layer thicknesses thereby increasing economic efficiency[115]. Further research activities centre on the application of multiple materials in the same process to enable functional coatings or geometry adjusted material properties. By using laser cladding (see Section3.3.6), the combination of different materials to produce functionally graded parts was proven to be feasible for Ni-base alloys and Ti-based alloys by Domack and Baughman [58]. With SLM, the manufacture of metallurgically bonded joints between WC/Co and steel for use in tool construction has been demonstrated[146] and [183]. Functionally graded zirconia Ni-based alloy parts on Waspaloy substrates have successfully been built [140]. Here, the mass fraction of Zr in the powder mixture increases layer by layer with build height. This layer-wise change of powder composition limits graded material transitions in one direction. By using selective coating mechanisms, it is also possible to create functionally graded materials in any desired orientation [146] and [147].

Mechanical Properties and Microstructure

The mechanical properties and microstructural features of additively manufactured components are governed by the high cooling rates during the solidification process, normally resulting in small grain sizes. Since Inconel 718 is an age-hardenable austenitic material, its strength is largely dependent on the precipitation of the γ⃞-phase. By appropriate heat treatments, the mechanical strength becomes comparable to wrought material.

In the as-fabricated form, the material has non-isotropic strength and strain properties. Upon hot isostatic pressing the microidentation hardness (HV) can be increased to approximately 6 GPa while the recrystallization can reach up to 10 vol% [7]. The precipitation strengthening of γ⃞ and γ⃞ and the inter-crystalline strengthening of needle-like γ-phase at the grain boundaries are the main reasons for the high micro-hardness values seen in the Inconel 718 samples fabricated by SLM and post age-hardening treatment [201]. The level of microstructural defects (porosity, junction errors, etc.) can be decreased by moving towards higher energy inputs (lower layer thickness, lower scanning speed and higher laser power) and is typically in the order of 0.01%. Additionally, advanced technologies such as Selective Laser Erosion and Laser Remelting to improve surface integrity are under development [218].

The right hand photograph in Fig. 47 shows that the microstructure of Ti–6Al–4V processed by laser-based additive manufacturing is mainly composed of acicular martensite (α⃞ phase). While the yield stress and the ultimate tensile strength are relatively high, the ductility can be considered as low (<10%). Post-process heat treatment can improve the mechanical properties of the material by changing the microstructure and reduce thermally induced residual stresses in the built part. Due to the fact that the resulting microstructure differs significantly from bulk Ti–6Al–4V alloy, other heat treatment parameters compared to forged or cast Ti–6Al–4V are necessary [197].

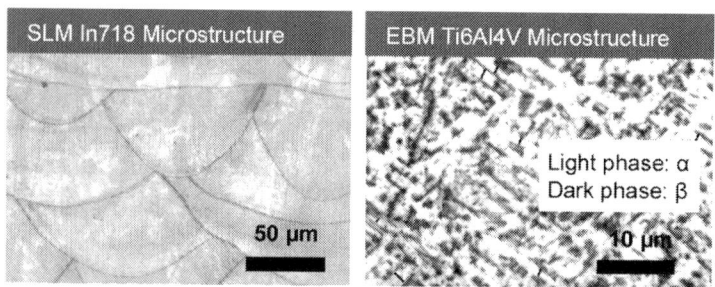

Figure.47: Microstructure of as-fabricated In718 showing consolidation structure (courtesy of IWB) and of EBM-processed Ti6Al4V [90].

Electron beam melted Ti–6Al–4V exhibits anisotropic behaviour of its mechanical properties according to studies performed by Hrabe and Quinn [90] and [91]. Elongation in the vertical direction (parallel to the electron beam) is detailed as ~30% lower than is the case for horizontally oriented parts. In contrast to classical annealing where the microstructure is typically discrete α in continuous β, the microstructure obtained by EBM was found to be continuous α with discrete β.

Other materials like TiAl are also of interest for future jet engine applications because of their low mass compared to nickel-based alloys. While EBM produces parts with a relative density of approximately 98% (above 99% after hot isostatic pressing), TiAl components manufactured with SLM show poor mechanical properties and a relative density of ~97% due to cracks [197]. For Ti–6Al–4V densities up to 99.98% can even be achieved as-built. Further microstructural analysis of standard as well as advanced alloys such as Ti–6Al–4V [41] and [220], Inconel 718 [184], Rene41 [127] and TiAl [141] are detailed in several publications.

Quality Methods

Currently, post-build inspection procedures account for as much as 25% of the time required to produce an additive manufactured engine component. However, by conducting inspection procedures during component build-up, a significant acceleration of production rates is expected [70]. Common post-build quality methods in aeroengine applications include, among others, the verification of part geometry through noncontact optical measurements. Since geometrical accuracy is crucial for functionality, a great deal of effort has been put into reliable determination of the entire external geometry. Testing for inner defects is done through well-established non-destructive methods like X-ray or ultrasonic inspection.

Taking advantage of the layer-wise build up procedure in additive manufacturing, online monitoring can provide complete part inspection during manufacturing. The principal approaches can be divided into coaxial setups sensing the process emissions directly at the current beam position with off-axis setups usually monitoring the complete build substrate at regular intervals.

A coaxial setup, which focuses on monitoring the irradiance emitted by the melt pool was developed by KU Leuven [46]. Since the system uses the same scanning unit for material processing and process monitoring, the detector elements are always focused on the current process zone. The total melt pool area is identified to be the relevant detection variable when analyzing process irregularities. With regard to real-time process control, commercial implementation of this system working in the kHz range is currently under development [23]. The usable wavelength band is however severely restricted to a small band around the laser wavelength because the same optics have to be employed.

The feasibility of layer-wise process monitoring based on a micro-bolometer IR-camera is shown for SLM by Krauss et al. [114]. Deviations in the laser melting process, occurring at a timescale of several tens of milliseconds can be detected by evaluation of properties of the heat affected zone under idealised conditions. Thermographic process surveillance, see Fig. 48, is particularly well suited for investigating the heat balance and geometry dependent heat accumulation resulting in inhomogeneous material properties and residual stresses. In this regard, the understanding of mechanisms of heat regulation [225] is of fundamental importance. For the EBM process, an IR camera based approach has been developed[162] and [169] focusing on the flaw detection directly after layer solidification. To overcome the shortcoming of not being able to monitor the actual solidification process due to metallization effects, Dinwiddie et al. [57] analysed different materials for use as a continuously rolling window for the EBM process. Further process monitoring approaches are detailed in several other publications [27] and [40].

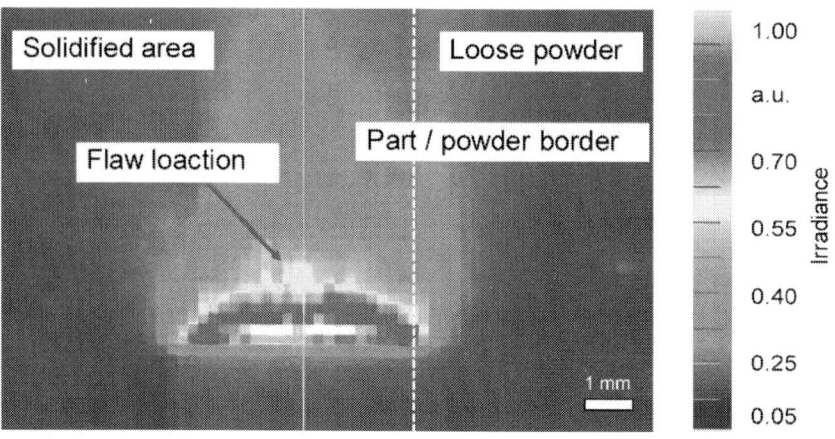

Figure.48: Thermogram of the heat affected zone after passing of an artificial flaw during SLM [114].

Repair Cladding

The efficiency of machinery is directly dependent on the degree of wear and/or erosion, which is an unavoidable result of usage. Worn turbine blades can affect the whole performance of the system and cause a reduction of the turbine efficiency producing high energy losses. Even minor damage on the tips of turbine blades can have major implications. In order to maintain a high performance level, excessively worn turbine blades are either replaced or repaired. As these turbomachinery components are made of high quality and therefore expensive Ni- or Ti-based alloys [156], the repair of these parts is often first choice.

Laser cladding by powder injection is one possible technology to restore for example the tips of worn turbine blades. Such laser based technology is known by various names e.g. laser metal deposition (LMD), laser engineered net shaping (LENSTM) or direct metal deposition (DMDTM). The LENSTM technology was developed at the Sandia National Laboratories in 1995 [126]. By focussing a high power solid-state laser (in general fibre or disc laser) onto a metal surface, a melt pool evolves as a consequence of absorbed laser light. From a powder nozzle that is either attached laterally or coaxially to the focussing optics, metal powder is injected into the melt pool.

This instantly changes to a liquid phase after reaching the melting temperature. Because of the moving laser focus, a movement that is determined by the feed direction and feeding speed, the liquid metal re-solidifies and as a result a weld track evolves on the workpiece surface. For protection and transport, a chemical non-reactive inert gas such as argon or helium is applied to prevent oxidation of the metal powder during the melting process. A detailed description of the laser metal deposition process and system technology can be found in several texts [132] and [190]. In Fig. 49, a schematic of the basic process setup of laser cladding by powder injection and an application example involving edge repair is displayed.

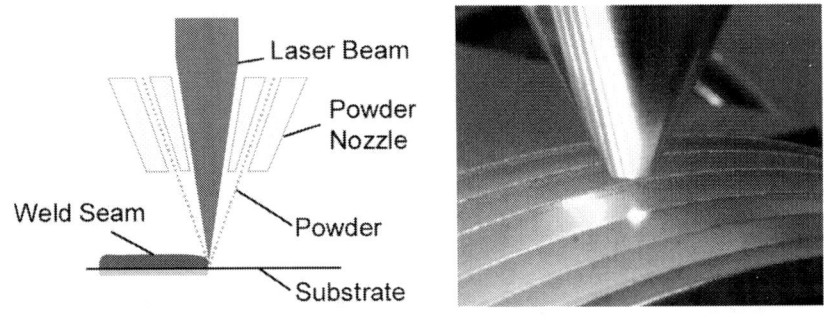

Figure.49: Basic setup of the laser metal deposition (LMD) process, based on [18] and example of application during edge repair [69].

As described previously, LMD is an economical and highly flexible repair technique which is already applied in the turbomachinery industry [26], for example to build-up worn tips of turbine blades and labyrinth seals [18] and [161], see Fig. 50. Further examples are the reconditioning of groove walls on HPC front drums made of Ti–6Al–4V and Ti-6246 or the repair of flanges on HPT produced with Nimonic PE16 with Inconel 625 powder. Due to the minimal heat input from the LMD repair process, distortion of the component was almost completely avoided [103]. Process adaptations have been developed especially for the repair of Ni-based superalloys with high Ti/Al composition in order to reduce cracking susceptibility [26]. For the repair of aeroengine components the Fraunhofer ILT in Germany for example has already been successfully certified by Rolls-Royce Deutschland for 15 different repair applications [68].

Figure.50: Blading repair (red boxes) with laser cladding: Turbocharger impeller [99], gas turbine blade [69] and labyrinth seal features [159].

In general, different approaches are possible for the repair of worn turbine blades, however coaxial powder nozzles have proved to be more suitable for this task than lateral arrangements [182]. The major disadvantage of a laterally positioned nozzle is the fact that the three-dimensional free moving space is limited and hence, the flexibility of the process is reduced. Furthermore, the quality of the welding tracks produced with lateral nozzles strongly depends on the feeding rate direction, whereas those fabricated by applying a coaxial powder nozzle are not directly influenced by the feeding rate direction.

Although coaxial powder nozzles are well-established in the field of turbine blade repair, powder efficiency falls behind that of a lateral powder nozzle [182]. As the materials of a turbine blade are precious, different approaches have been developed to increase the powder efficiency of the repair process. For example copper moulds in which the tip of the blade is re-built up are used [182]. Here, not only material waste can be avoided but also the required time for post-processing such as machining can be significantly reduced. Furthermore, the time for the repair process can be decreased by applying special optics. In order to produce weld seams of an aerofoil, laser focusing optics (generally referred to as Zoom optics) that can dynamically change the focus diameter during the process have been developed.

The laser cladding process is also an established technique for coating applications. It has been used to generate wear-resistant coatings on the shroud shelves of turbine engine blades in order to increase the wear resistance and lifetime of these components [174]. Additionally, resistance against wear of certain part sections can be achieved with hard-facing by the use of laser cladding [159]. In this context ceramic materials such as zirconium are applied to deposit a wear resistant protective layer.

Post Processing and Finishing of AM Components

In laser cladding additive manufacturing, only near net shape structures can be generated or restored, meaning that post-processing, primary milling to achieve the desired surface quality and the final part geometry is always required. Hence, the combination of laser cladding and machining is part of current research activities. For example since 2008, a UK based consortium has been developing a combined approach known as the "Remanufacture of high value products using a combined Laser cladding, Inspection and Machining (RECLAIM)" system [99]. Within the Fraunhofer Innovation Cluster "TurPro – Integrative Produktionstechnik für energieeffiziente Turbomaschinen" in Germany, similar repair as well as new part production process chains have been developed including a complete CAx framework approach[69]. Fig. 51 shows typical examples of subsequent post processing of AM manufactured parts in order to achieve the final functional macro and micro workpiece geometry. Additionally, ECM technologies are also gaining interest as an alternative to conventional cutting and/or polishing or blasting operations.

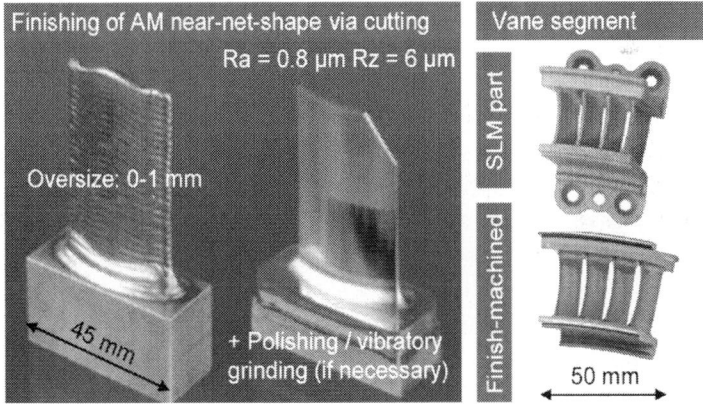

Figure.51: Examples for subsequent post processing of functional surfaces on AM manufactured components: Near-net-shape HPC In718 blade mock-up [69] and nickel-based vane segment [159].

Photonic Processes Ii: Laser Drilling of Cooling Holes

Introduction

Approximately 5% of all industrial laser material processing applications are laser drilling operations [25]. In this context, the generation of cooling holes in gas turbines for aircraft as well as for power plants is one of the most important, established drilling applications. The steady progression of laser-based system technology (e.g. laser sources, data preparation software, machine control, positioning system, sensor devices, etc.) and the development of novel laser drilling strategies, offering design freedom together with its cost effectiveness has increased significantly during the last decade. Assuming that suitable technology for industrial series production is available, it is possible to drill hundreds/thousands of cooling holes with high precision and of variable diameter and shape in multi-material blades of complex geometry, see Fig. 52.

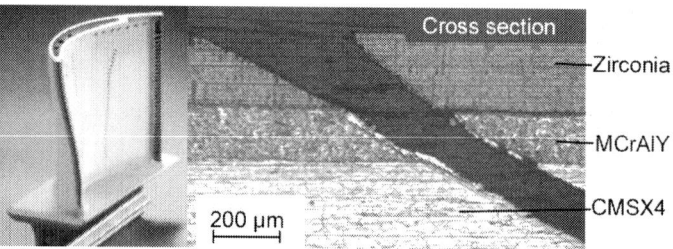

Figure.52: Laser drilling of cooling holes: Nickel-based turbine blade coated with a ceramic wear resistance layer with cooling holes [25] and cross-section of a cooling hole drilled in a CMSX-4 turbine blade coated with a MCrAlY and zirconia wear resistance layer [89].

One of the main challenges is drilling through multi-layer material systems composed of metal coatings with at least one ceramic wear protection layer without enhanced formation of micro-cracks and thermal induced damage causing a removal of the coating layers. Furthermore, to achieve significant cooling performance the cooling holes are generally tapered or shaped.

Theoretical Background

In general pulsed laser systems are applied for laser drilling processes, where selection of the pulse duration depends on the hole characteristics and the material being processed. In the field of laser drilling of turbine components, pulse duration is normally of the order of nano- or milliseconds. The required pulse energy basically depends on the exact chemical composition of the material, the material thickness and the desired hole diameter and shape. In this context an increase of pulse energy generally causes higher drilling rates, but a decrease of drilling quality such as enhanced formation of melting deposits on the workpiece surface and hole edges.

The fundamental interaction mechanisms during laser drilling are schematically illustrated in Fig. 53. A detailed explanation of the different physical mechanisms is detailed by Poprawe [153]. Different physical mechanisms take place inside the irradiated material volume depending on laser pulse duration and laser pulse peak intensity. The dominating effects causing material removal are melting and vaporisation. Assuming a temporal and spatial Gaussian intensity distribution of the incoming pulsed laser beam, vaporisation occurs in the hole-centre and melting in surrounding material sections. For a laser peak intensity of approximately 10^6 W/cm^2 and pulse durations of the order of several milliseconds, melting is the dominating effect being responsible for material removal and hole formation.

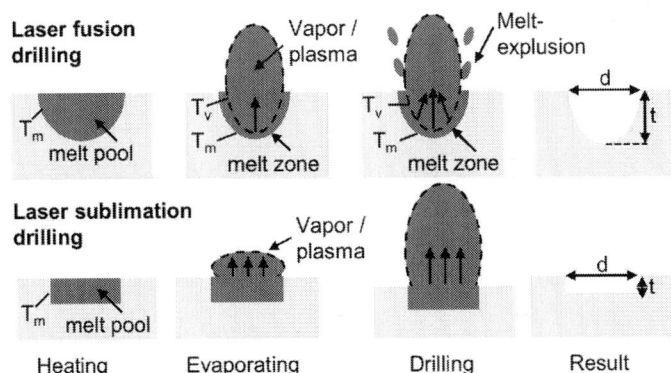

Figure.53: Physical interaction mechanisms during laser fusion drilling and laser sublimation drilling. Based on [153].

Initially, the irradiated material is heated up to its melting temperature and a significant melt pool evolves. As a consequence of further energy input vaporisation occurs. Along the drilling axis a dense vapour plume is formed that cannot pass the molten material. Due to recoil of the vapour plume, the melting at the bottom of the hole is continually accelerated and the molten metallic material is expelled along the hole edges. The expulsion of molten vapour material is assisted by applying an inert gas process-stream. Although typically coexisting with vaporisation, the dominating effect is melting and the process causing material removal is melt expulsion. In this context, the drilling process is denoted as fusion or heat drilling. As a consequence of melting, expulsion of small re-solidified particles remain at the hole edges. To achieve typical cooling hole depths of between 8 and 25 mm, a sequence of several laser pulses are required. Furthermore, additional post-irradiation is necessary to remove the re-solidified particles from the edges of the holes and to achieve improved hole quality.

With a further increase of laser peak intensity greater than 10^6 W/cm^2, sublimation drilling occurs. In this case, the dominating effect causing material removal is ablation by vaporising plasma formation. In this context, the applied peak intensity has to exceed a material-dependent threshold value. To achieve such high intensities, shorter laser pulses are required. A system suitable for sublimation-drilling is a Q-switched Nd:YAG laser with pulse durations of the order of 10–100 ns A detailed explanation of plasma formation and resulting material removal can be found in the literature [132] and [153].

Drilling Operations

Industrially established laser drilling operations include single-pulse drilling, percussion drilling, trepanning and helical drilling. For cooling holes in turbine blades the relevant drilling operations are percussion drilling and trepanning [25]. Furthermore, there exist different approaches to drilling holes of complex shape and varying diameters by combining both. The principles of trepanning and percussion drilling are schematically shown in Fig. 54. A detailed theoretical description of the drilling technologies is presented by Poprawe [153] and Majundar [132]. During the drilling process, an inert process gas stream protects the focusing optics. The gas stream is also employed to assist material removal and to prevent oxidation and necking of the holes by melting deposits from ablated material.

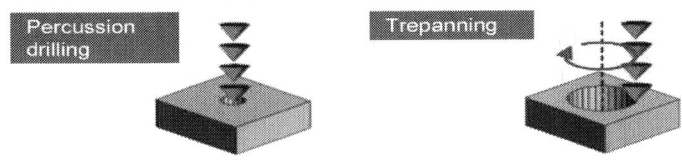

Figure.54: Principle of percussion drilling and trepanning, [52].

During percussion drilling the laser spot is stationary at the same position on the workpiece in contrast to the trepanning operation. The diameters of the approximately cylindrical holes are commonly of the order 0.5–0.7 mm and the achievable aspect ratio is in the best case 1:20 [25]. An important point is the exact positioning of the focus plane relative to the workpiece surface, which is a function of the required results and workpiece material, the optimal position of the focus spot being located approximately 5–15% of the workpiece thickness under the workpiece surface [25]. In practice the best setting for a specific problem will be empirically identified by analysing hole quality in terms of geometry, distribution of cracks and the amount of ablated material recombining at the hole edges and on the workpiece surface [125].

When trepanning, the laser beam is rotated relative to the workpiece, whereby the laser spot diameter is distinctly smaller than the diameter of the hole. A hole is generated by removing a cylindrical core during one circulation of the focused laser beam. Technically speaking trepanning is a laser-based hybrid process involving classical drilling and laser cutting. Additionally during laser spot circulation, the angle of incidence on the workpiece surface can be varied by influencing the taper angle of the drilled hole [24]. Tilting the focused laser beam is realised by an arrangement of rotating optical components, for example a dove prism with a defined beam displacement at the entrance of the optical element, or by means of tilting reflective optics [24]. The principle of a rotating laser spot provides the opportunity to generate holes with high reproducibility and high flexibility in terms of the hole design. For instance, the hole diameter is not directly related to the laser spot diameter as is the case for percussion drilling. Furthermore hole shape can be non-circular. In comparison to percussion drilling, trepanning is more time-consuming and the heat input into the turbine blade is larger.

Generation of Complex Cooling Holes Via Process Combination

Two types of cooling holes are used for turbine blades, one shaped like conical nozzles the other cylindrical holes where only the exits possess a conical shape. This latter type affords an improved cooling performance based on effusion cooling [154]. In this case a thin film of cool air originates above the aerofoil above the hole edges, resulting in an increase of thermal shielding [25]. This significantly increases turbine entrance temperatures giving higher overall efficiency and reduced fuel consumption. In the past, such cooling holes were exclusively fabricated by EDM however, laser drilling is increasingly being used due to its higher productivity (shortest machining times), superior production line capability and greater flexibility. While EDM has comparatively lower machine tool costs, parallelised machining capability [187] and better process control in terms of break through detection as well as reduced heat affected zones, it is limited to the machining of electrically conductive materials and therefore cannot be applied for ceramic-coated turbine blades.

A method to drill cylindrical cooling holes with conical shaped exits in ceramic-coated turbine blades is the combined use of laser percussion drilling and trepanning operations by applying different laser sources. In this case first a cylindrical through-hole is produced by percussion drilling with a lamp-pumped Nd:YAG-laser, before the shaped hole exit is formed by trepanning. This procedure as suggested by Beck [25] is schematically shown in Fig. 55 together with an example of application. A Q-switched Nd:YAG laser system with laser pulses of the order of 100 ns pulse duration and high peak powers between 10 and 100 MW/cm^2 are applied for the trepanning setup. In contrast to flash-lamp pumped Nd:YAG solid state lasers with pulse durations of 10–100 µs, the dominating effect causing material removal is ablation and vaporisation of the irradiated material volume. With regard to cooling holes in turbine blades, the combined use affords the possibility to switch dynamically between ablation and melt expulsion during hole formation. This procedure allows the generation of cooling holes with maximum flexibility in terms of the shape, design and drilling depth for improved cooling performance. Additional research has focussed on achieving controlled melt flow by utilisation of a secondary gas jet in order to avoid delamination effects during machining of coated blades [172]. Other process combinations such as sequential laser and mechanical drilling [145], combination of laser and EDM [193] and even EDM with mechanical drilling [67] are also currently under development.

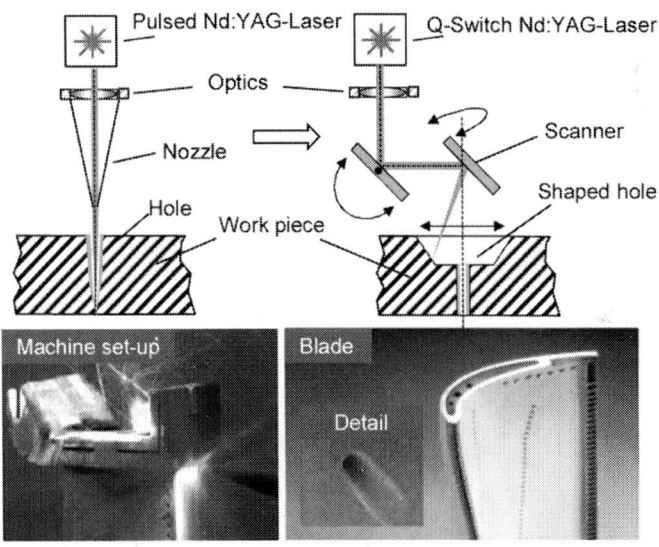

Figure.55: Principle for process combination of cylindrical drilling and shape hole drilling in one machine set-up and example of high pressure turbine blade with shaped holes. Based on [25] and [159].

ECONOMICAL PROCESS CHAIN ANALYSIS AND COST MODELLING

For efficient turbomachinery component manufacture, an economic analysis of individual process technology alternatives as well as the resulting process chains is required. Machining processes with low material removal rates but relatively low machine running costs such as EDM must be evaluated in such a way in order to be competitive against other conventional or advanced manufacturing technologies. Therefore, by performing parallel technological and economic analysis of each operation, optimal processes and process chains can be identified to minimise production costs.

As an example, Fig. 56 shows possible competing machining technologies for the manufacture of a blisk with "hyper-polished" blading. The technologies shown have provisionally been identified as possible variants/alternatives from a technological point of view. While additive manufacturing with complete build-up of blades and water-jet machining from solid are able to produce a rough blade contour, subsequent multi-axis milling could be used for finish machining or for the initial roughing, with conventional or ECM-based polishing applied to produce the finished blades. Alternatively, ECM and PECM (with high MRR) could be employed for the whole process, as they are capable of roughing, finishing and polishing with the same base technology. While all other process technologies need separate machine tools, ECM and PECM can be realised in one set-up. Conversely, such machine tools together with the process and tool electrode design are complex, and therefore more expensive. In order to identify an economic alternative, an appropriate cost model considering all relevant boundary conditions has to be executed and constantly updated when key parameters are changing.

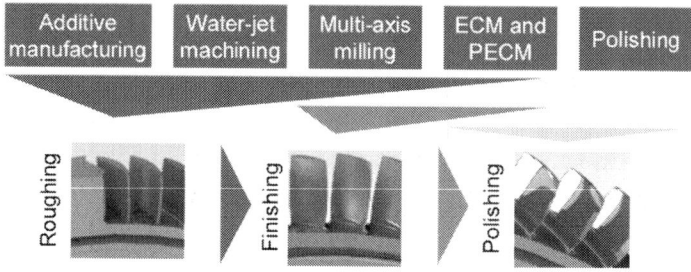

Figure.56: Example of competing machining technologies for blisk manufacturing process chain of roughing, finishing and polishing.

Such an economical comparison has been carried out for different roughing strategies for blisk gap slotting from solid via multi-axis milling, sinking-EDM and ECM for titanium- and nickel-based blisks. The analysis used a characteristic geometry and the technological key parameters detailed in Fig. 57, the resulting manufacturing costs being calculated as a function of batch size.

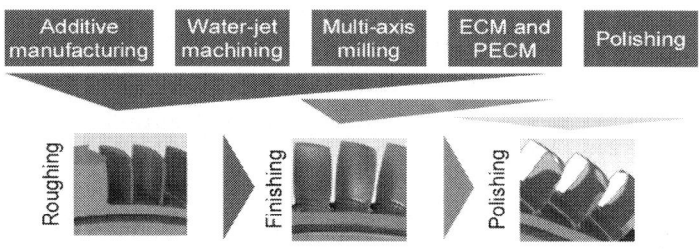

Figure.57: Simplified blisk gap geometry and averaged material removal rates derived from technological analysis, [111].

Material removal rates have the greatest influence on overall manufacturing costs, as they affect the direct process time and are thus responsible for machine hourly rates and wage costs. In contrast, tooling costs have limited influence although they have to be taken into account. To allow a comparison, other manufacturing parameters appropriate to blisk manufacture have to be kept constant. These parameters and further assumptions are listed below and provide the basis for the final economic calculation. The idealised blisk possesses 36 blades. Investment/capital costs of machine tools were obtained as official quotations from manufacturers (specialised 5-axis milling machine for turbine industry 1000 k€, SEDM machine 190 k€, ECM machine 400 k€ (plus tooling costs), graphite milling machine 220 k€). Set-up time is equal for each manufacturing technology and amounts to 1 h. Wage costs amount to 35 €/h. For each blisk made of Ti–6Al–4V one milling tool wears out while for Inconel 718 two cutters are required (each at 300 €). Manufacturing and material costs have been considered for SEDM electrodes. The ECM tool operates in the axial direction (marked area). One employee can operate one milling- or ECM machine or alternatively three EDM machines at once. Milling- and ECM machines are used in a two shift production (3200 h/a) and EDM machines in a three shift arrangement (4800 h/a, unmanned EDM production is state of the art) [113].

In Fig. 58, solutions for Ti–6Al–4V and Inconel 718 are shown. Steps in the responses represent critical batch sizes for which full capacity utilisation is reached. Here, the roughing costs per blisk are at a minimum for the respective technology. One-off tool development costs in ECM can be difficult to predict. For this reason, two envelope curves are drawn which consider a lower limit of 100 k for tool development costs and an upper limit of 500 k . Except in the case of small batch sizes from 0 to 20 blisks per year, rough machining via SEDM of Ti–6Al–4V would be simply uneconomic. Milling from solid and ECM – low tool developing costs assumed – are the two technologies in competition. This means that for Ti-based blisks, depending on specific geometries, a closer economic analysis has to be made. For Inconel 718 the choice of the most economical roughing process is more difficult. With low machine tool investment costs and relatively high average material removal rates, SEDM especially for batch sizes up to 400 blisks per year, is a viable alternative. In the case of larger batch sizes, ECM is the most cost-effective technology. Batch sizes of 100–300 require a more detailed analysis [113].

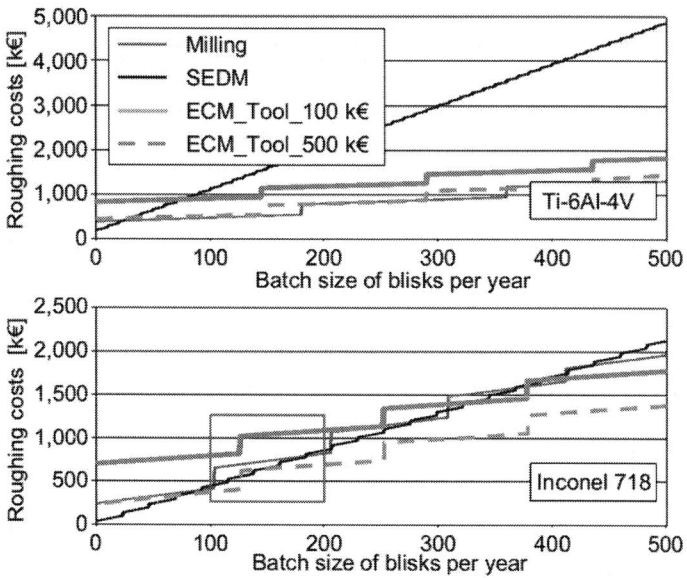

Figure.58: Roughing costs as a function of batch size for Ti–6Al–4V and In-conel 718 [113].

Roughing costs per blisk (Inconel 718) are shown in Fig. 59. Here, costs strongly depend on the batch size as well as investment levels where additional machine tools are required. Technologies like SEDM with low investment costs allow less volatile curve progressions but machine tools and therefore space, is needed more frequently. In the example for a batch size of 200, milling, SEDM and ECM with low tooling outlays reach the same cost level. The local minima of roughing costs – caused by full capacity utilisation – were approximately equivalent. With higher numbers of machine tools, the maximum roughing costs per blisk decrease. This effect is due to the increase in capacity utilisation for each machine tool so that single investment costs are normalised [113].

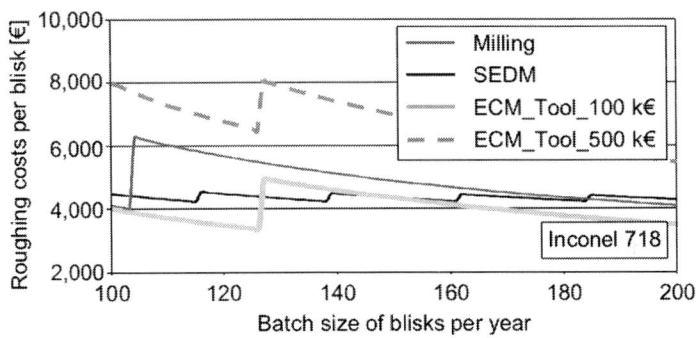

Figure.59: Roughing costs per blisk as a function of batch size for Inconel 718 (detail according to red box of Fig. 57), [113].

A similar economic analysis will also have to be carried out for the evaluation of WEDM as a possible alternative technology for fir tree production (assuming proof and acceptance of resulting workpiece functionality) [206]. Although EDM machining time is 10 times longer compared to the established broaching process, the overall costs are estimated to be of the same order of magnitude when taking machine investment costs, tooling and further operating costs into account. Furthermore, WEDM offers higher machining flexibility as well as significantly reduced lead times for new tools and/or new designs.

In addition to the aforementioned characteristics of single operations, further aspects of the complete process chain must be taken into account during technology evaluation. During application of ECM for example, no additional de-burring operation is required in the process chain, as workpiece edges will automatically be rounded. On the other hand, additional washing operations may be necessary. Energy and recycling costs are similarly gaining importance and therefore have to be considered. Recycling of contaminated chips from conventional cutting operations will also need to be critically evaluated [188], as do scrap blocks cut out via WEDM operations. The residual tensile stresses from EDM material removal could be neutralised by subsequent surface finishing operations such as etching, vibro/shot peening, vibratory grinding or abrasive flow machining. Thus, in order to substitute established conventional manufacturing technologies and process chains a comprehensive evaluation of Low Cycle Fatigue (LCF), High Cycle Fatigue (HCF) and Thermal Mechanical Fatigue (TMF) – depending on the chosen turbomachinery component – has to take place relative to the established processes and in dependence on the specific surface integrities. Finally, by using a non-mechanical material removal process, induced forces become less significant, allowing the machining and therefore the design of more filigree and complex geometries, which to date have not been possible by utilising conventional means.

For a comprehensive and balanced economic evaluation of additive manufacturing technologies, novel/new cost model approaches with an extended frame of examination are currently under development. These include the assessment of primary process costs employed to manufacture semi-finished parts for the subtractive process chains as well as consumables, energy and also recycling costs which have up to now, generally not been taken into consideration.

SUMMARY AND CONCLUSIONS

The technical capabilities and areas of application of electro-chemical, electro-physical and photonic processes have been analysed showing the broad potential of ECM, EDM, additive manufacturing and laser material removal for the manufacture of turbomachinery components. Clear advantages have been identified for their use when machining advanced and difficult-to-cut materials, including high removal and deposition rates, superior geometrical precision and acceptable surface integrity. The case for their selection as alternatives to conventional cutting technologies is further strengthened when considering economic aspects, as outlined in the example for roughing of blisk gaps.

ACKNOWLEDGEMENTS

The authors would like to express their sincere thanks to R. Perez (GF AgieCharmilles), M. Cuttell (Rolls-Royce), P. Harpham, J. Durrant (Siemens Industrial Turbomachinery); M. Zeis, D. Welling (WZL RWTH Aachen University); H. Krauss, J. Weirather (IWB, TUM) and O. Hentschel, M. Karg, C. Scheitler (LPT, Universität Erlangen) for their assistance in the paper preparation. We would also like to thank the STC-E members involved for their help.

REFERENCE

1. Aas KL (2004) Performance of Two Graphite Electrode Qualities in EDM of Seal Slots in a Jet Engine Turbine Vane. Journal of Materials Processing Technology 149:152–156.

2. ACARE (2014) Aviation in Europe: A vision for 2050. www.acare4europe.org.

3. Additive Manufacturing at GE (2011) Loughborough. July 12.

4. Ali S, Hinduja S, Atkinson J, Pandya M (2009) Shaped Tube Electrochemical Drilling of Good Quality Holes. CIRP Annals – Manufacturing Technology 58:185–188.

5. Alkahari MR, Furumoto T, Ueda T, Hosokawa A, Tanaka R, Aziz A, Sanusi M (2012) Thermal Conductivity of Metal Powder and Consolidated Material Fabricated Via Selective Laser Melting. KEM 523–524:244–249.

6. Allen J (2011) The Potential for Aero Engine Component Manufacture Using Additive Layer Manufacturing. Aerodays Conference, Madrid.

7. Amato K, Gaytan S, Murr L, Martinez E, Shindo P, Hernandez J, Collins S, Medina F (2012) Microstructures and Mechanical Behavior of Inconel 718 Fabricated by Selective Laser Melting. Acta Materialia 60(5):2229–2239.

8. Antar MT, Soo SL, Aspinwall DK, Cuttell M, Perez R (2009) Machined Workpiece Surface Integrity Effects Using Advanced Wire EDM. Proceedings of the 26th International Manufacturing Conference (IMC26), Trinity College Dublin, Ireland, 103–110.

9. Antar MT, Soo SL, Aspinwall DK, Cuttell M, Perez R, Winn AJ (2010) WEDM of Aerospace Alloys Using 'Clean Cut' Generator Technology. Proceedings of the 16th International Symposium on Electromachining (ISEM XVI), 19–23 April, Shanghai, China, 285–290.

10. Antar MT, Soo SL, Aspinwall DK, Jones D, Perez R (2011) Productivity and Workpiece Surface Integrity when WEDM Aerospace Alloys Using Coated Wires. Procedia Engineering 19:3–8.

11. Antar MT, Soo SL, Aspinwall DK, Sage C, Cuttell M, Perez R, Winn AJ (2012) Fatigue Response of Udimet 720 Following Minimum Damage Wire Electrical Discharge Machining. Materials and Design 42:295–300.

12. ASME Turbo Expo 2013 (2013) Turbine Technical Conference and Exposition. Monday, June.

13. Aspinwall DK, Dewes RC, Mantle AL (2005) The Machining of γ-TiAl Intermetallic Alloys. CIRP Annals – Manufacturing Technology 54(1):99–104.

14. Aspinwall DK, Soo SL, Berrisford AE, Walder G (2008) Workpiece Surface Roughness and Integrity After WEDM of Ti–6Al–4V and Inconel 718 Using Minimum Damage Generator Technology. Annals of CIRP 57(1):187–190.

15. Aspinwall DK, Soo SL, Curtis DT, Mantle AL (2007) Profiled Superabrasive Grinding Wheels for the Machining of a Nickel Based Superalloy. Annals of CIRP 56(1):335–338.

16. Ayesta I, Izquierdo B, Sanchez JA, Ramos JM, Plaza S, Pombo I, Ortega N, Bravo H, Fradejas R, Zamakona I (2013) Influence of EDM Parameters on Slot Machining in C1023 Aeronautical Alloy. Procedia CIRP 6:129–134.

17. Bache MR, Tuppen SJ, Voice WE, Lee HG, Aspinwall DK (2009) Novel Low Cost Procedure for Fabrication of Diffusion Bonds in Ti6/4. Materials Science and Technology 25(1):39–49.

18. Balazic M, Milfelner M, Kopac J (2010) Repair and Manufacture of High Performance Products for Medicine and Aviation with Laser Technology. AIJSTPME 3(4):15–22.

19. Baumga¨rtner M (2006) Entwicklung Innovativer ECM-Verfahren fu¨r die Herstellung von Leitschaufelsegmenten, Luftfahrtfoschungsprogr. LuFoIII.

20. Baumga¨rtner M (2010) Herstellung von Turbinenkomponenten, series of lectures "Ausgewa¨hlte Kapitel der Turbomaschinen", AKTM, RWTH Aachen University.

21. Baumga¨rtner M (2013) Entwicklung des elektrochemischen Senkens (ECM) und der mechanischen Bearbeitung von Titanaluminiden, Abschlussbericht BMBFForschungsvorhaben.

22. Bayer E (2014) Fertigungsprozesse als Schlu¨sseltechnologie zur Realisierung des Geared Turbo Fan. Retreived from www.mtu.de.

23. Bechmann F, Berumen S, Craeghs T, Clijsters S (2012) Prozessu¨berwachung und Qualita¨tssicherung generativ gefertigter Bauteile, Rapid. Tech, Erfurt.

24. Bechtold P, Mayer G, Lu¨hring P, Schmidt M (2012) Novel System Technology for Trepanning a Laser Beam at Arbitrary Developments of Angle of Incidence and Lateral Focus Position. Key Engineering Materials 523–524:238–243.

25. Beck T (2011) Laser Drilling in Gas Turbine Blades – Shaping of Holes in Ceramic and Metallic Coatings. Laser Technik Journal 3:40–43. (WILEY-VCH, Weinheim).

26. Bi G, Gasser A (2011) Restoration of Nickel-Base Turbine Blade Knife-edges with Controlled Laser Aided Additive Manufacturing. Physics Procedia 12:402–409.

27. Bi G, Schu¨rmann B, Gasser A, Wissenbach K, Poprawe R (2007) Development and Qualification of a Novel Laser-Cladding Head with Integrated Sensors. International Journal of Machine Tools & Manufacture 47:555–561.

28. Bilgi DJ, Jain VK, Shekhar R, Mehrotra S (2004) Electrochemical Deep Hole Drilling in Super Alloy for Turbine Application. Journal of Materials Processing Technology 149:445–452.

29. Boeing Current Market Outlook 2013-2032, (2013), . www.boeing.com/cmo.

30. Bogoveev NA, Firsov AG, Filatov EI, Tikhonov AS (2001) Computer Support for "All-Round" ECM Processing of Blades. Journal of Materials Processing Technology 109:324–326.

31. Bradley A (2010) Engine Design for the Environment, RAeS, Hamburg. June.

32. Broichhausen K (2004) Innovationsforum 'Flugtriebwerkstechnik in Brandenburg' Perspektiven fu¨r die Einbindung von KMU's

in die Triebwerksindustrie, Conference: Brandenburgische TU Cottbus 7–8. Dezember.

33. Burger M (2010) Prä"zise Elektrochemische Bearbeitung (PECM) komplexer Strukturbauteile, (Dissertation) TU Mu"nchen, Mu"nchen.

34. Burger M, Koll L, Werner EA, Platz A (2012) Electrochemical Machining Characteristics and Resulting Surface Quality of the Nickel-Base SingleCrystalline Material LEK94. Journal of Manufacturing Processes 14:62–70.

35. Burger M, Platz A, Werner E (2007) Herstellung/Nachbearbeitung von Turbinenblisks durch Prä"zises Elektrochemisches Bearbeiten, TUM. Retrieved from http://www.wkm.mw.tum.de/forschung/postergalerie/.

36. 36. Bußmann M, Bayer E (2013) Market-Oriented Blisk Manufacturing – A Challenge for Production Engineering, MTU Aero Engines.

37. Bußmann M, Kraus J, Bayer E (2005) An Integrated Cost-Effective Approach to Blisk Manufacturing. Retreived from www.mtu.de.

38. Bußmann M, Bayer E (2009) Blisk Production of the Future Technological and Logistical Aspects of Future-Oriented Construction and Manufacturing Processes of Integrally Bladed Rotors. Retrieved from www.mtu.de.

39. Chakradhar D, Venu Gopal A (2010) Design and Optimization of Process Parameters in Electrochemical Machining of Inconel 625 Alloy Using Taguchi Method. Proceedings of the 16th ISEM, Shanghai, China, 373–377.

40. Chivel Y, Smurov I (2010) On-Line Temperature Monitoring in Selective Laser Sintering/Melting. Physics Procedia 5(5, Part B):515–521.

41. Clark D, Whittaker MT, Bache MR (2012) Microstructural Characterization of a Prototype Titanium Alloy Structure Processed via Direct Laser Deposition (DLD). Metallurgical and Materials Transactions: B 43:388–396.

42. Clifton D, Mount AR, Jardine DJ, Roth R (2000) Electrochemical Machining of Gamma Titanium Aluminide Intermetallics. Journal of Materials Processing Technology 108:338–348.

43. Components for Compressors and Turbines, (2013), . Retrieved from www.leistritz.com.

44. Cooke A (2010) Variability in the Geometric Accuracy of Additively Manufactured Test Parts. Proceedings of The 21st Annual Solid Freeform Fabrication Symposium, An Additive Manufacturing Conference, Austin TX, 1–12.

45. Craeghs T, Bechmann F, Berumen S, Kruth JP (2010) Feedback Control of Layerwise Laser Melting Using Optical Sensors. Physics Procedia 5:505–514.

46. Craeghs T, Clijsters S, Yasa E, Kruth JP (2011) Online Quality Control of Selective Laser Melting. Solid Freeform Fabrication: Proceedings, University of Texas, Austin.

47. Crosby HC (1985) Wire-Cut and Vertical CNC EDM Systems: Merging Technologies. Proceedings of the Conference on Non Traditional Machining, December 2–3, Cincinnati, USA, 43–51.

48. Curtis DT, Soo SL, Aspinwall DK, Huber C, Fuhlendorf J, Grimm A (2008) Production of Complex Blade Mounting Slots in Turbine Disks Using Novel Machining Techniques. Proceedings of the 3rd International CIRP High Performance Cutting Conference, Vol. 1, Dublin, Ireland, 219–228.

49. Curtis DT, Soo SL, Aspinwall DK, Sage C (2009) Electrochemical Superabrasive Machining of a Nickel-Based Aeroengine Alloy using Mounted Grinding Points. CIRP Annals – Manufacturing Technology 58(1):173–176.

50. D'Amario R (2008) Method and Apparatus for Generating Machining Pulses for Electrical Discharge Machining, European Patent EP 1719570.

51. Dausinger F, Lichtner F, Lubatschowski H (2004) Femtosecond Technology for Technical and Medical Applications. Topics in Apllied Physics, 96 Springer. ISBN: 978-3-540-20114-4.

52. Dilba D (2011) Schnelle Schichtarbeit. MTU Report 2011(1):20–25. http:// www.mtu.de/de/take-off/report/archive/2011_1.pdf.

53. Dilba D (2012) Ho¨chste Pra¨zision. Report MTU Aero Engines. Retrieved from www.mtu.de.

54. Ding S, Yuan R, Li Z, Wang K (2006) CNC Electrical Discharge Rough Machining of Turbine Blades. Proceedings of the IMechE Part B: Journal of Engineering Manufacture 220(7):1027–1034.

55. Dinwiddie RB, Dehoff RR, Lloyd PD, Lowe LE, Ulrich JB, Stockton GR, Colbert FB (2013) Thermographic In-Situ Process Monitoring

of the Electron-Beam Melting Technology Used in Additive Manufacturing, SPIE DSS: 87050K.

56. Domack MS, Baughman JM (2005) Development of Nickel-Titanium Graded Composition Components. Rapid Prototyping Journal 11(1):41–51.

57. ECM/PECM Technologie Polieren   3D-Konturen, EMAG ECM.

58. Electrochemical Machining (2013) Retrieved from http://www.koepperninternational.com

59. Electrochemical Machining (ECM) (2014) Leistritz Turbomaschinen Technik. Retrieved from www.leistritz.com.

60. Elektrochemisches Abtragen (2009) Verein DeutscherIngenieure VDI-Richtlinie 3401-1(Entwurf) .

61. EMAG's ECM/PECM Machines (2013) Retrieved from http://www.onlineamd.com/Article.aspx?article_id=132602.

62. FAA Continuous Lower Energy, Emissions & Noise (CLEEN) Technologies (2012) Rolls-Royce Programm Overview. November 8, www.Rolls-royce.com.

63. Fleischer J (2011) Erodierbohren – Neue Wege und Anwendungsbeispiele. Fachtagung FunkenerosionWZL RWTH Aachen University, Aachen.

64. Fonda P, Wang Z, Yamazaki K, Akutsu Y (2008) A Fundamental Study on Ti– 6Al–4V's Thermal and Electrical Properties and Their Relation to EDM Productivity. Journal of Materials Processing Technology 202:583–589.

65. 67. Garzon M (2011) Integ-Micro Project: New High-Precision Micro-Machining Technologies, EMO Hannover, Germany.

66. Gasser A, Backes G, Kelbassa I, Weisheit A, Wissenbach K (2010) Laser Additive Manufacturing: Laser Metal Deposition (LMD) and Selective Laser Melting (SLM) in Turbo-Engine Applications. Laser Technik Journal 7(2):58–63. (WILEY-VCH).

67. Gasser A, Kelbassa I, Wissenbach K, Witzel J, Go¨bel M (2012) Additive ManuFacturing in Turbo-Engine Applications, AKL/EU Innovation Forum, Aachen.

68. GE Aviation (2013) GE Aviation Signs Additive Manufacturing Cooperative Agreement with Sigma Labs. Press Release, GE Aviation, Evendale, Ohio.

69. GE Aviation (2013) GE Aviation Signs Additive Manufacturing Cooperative Agreement with Sigma Labs.

70. Giese C (2005) Verfahrensvergleich EDM/ECM im industriellen Umfeld – Anwendungsgebiete von ECM, Fachtagung Funkenerosion, RWTH Aachen.

71. GKN Aerospace Capabilities (2010) Retrieved from www.gknaerospace.com.

72. 74. Gmelin T (2013) PECM – Anwendungsmo¨glichkeiten und Grenzen im Ver-gleich zur Funkenerosion. 9. Fachtagung Funkenerosion, WZL, RWTH, Aachen.

73. 75. Gong X, Anderson T, Chou K (2012) Review on Powder-Based Electron Beam Additive Manufacturing Technology. ASME/ISCIE International Symposium on Flexible Automation, 507–515.

74. Guo ZN, Lee TC, Yue TM, Lau WS (1997) A Study of Ultrasonic-Aided Wire Electrical Discharge Machining. Journal of Materials Processing Technology 63:823–828.

75. Han F, Wachi S, Kunieda M (2004) Improvement of Machining Characteristics of Micro-EDM Using Transistor Type Isopulse Generator and Servo Feed Control. Precision Engineering 28:378–385.

76. Han GC, Soo SL, Aspinwall DK, Bhaduri D (2013) Research on the Ultrasonic Assisted WEDM of Ti–6Al–4V. Advanced Materials Research 797:315–319.

77. Hascalık A, Caydas U (2007) A Comparative Study of Surface Integrity of Ti– 6Al–4V Alloy Machined by EDM and AECG. Journal of Materials Processing Technology 190:173–180.

78. Hascalik A, Caydas U (2007) Electrical Discharge Machining of Titanium Alloy (Ti–6Al–4V). Applied Surface Science 253:9007–9016.

79. Henne J (2013) Gearing Up for High Volume GTFTM Production. 21st ISABE Conference, Busan, Korea.

80. Hess T (2012) Aero Engine Parts Made of SLM – Status Quo and Future Potential. International Laser Technology Congress AKL'12 Aachen.

81. Heuer J, Fili W (2006) Konstruieren ohne fertigungstechnisches Limit. Konstruktion Zeitschrift fu¨r Produktentwicklung und Ingenieur-Werkstoffe, 4.

82. Hewidy MS, El-Taweel TA, El-Safty MF (2005) Modelling the Machining Parameters of Wire Electrical Discharge Machining of Inconel 601 Using RSM. Journal of Materials Processing Technology 169:328–336.

83. High-Tech Development and Manufacturing for Aero and Industrial Gas Turbine Components, (2010), Sulzer ELDIM.

84. High-tech made by MTU (2011) e-Paper. Retrieved from: www.mtu.de

85. Hinduja S, Kunieda M (2013) Modelling of ECM and EDM Processes. CIRP Annals – Manufacturing Technology 62(2):775–797.

86. Ho KH, Newman ST (2003) State of the Art Electrical Discharge Machining (EDM). International Journal of Machine Tools and Manufacture 43(13):1287–1300.

87. Horn A, Weichenhain R, Albrecht S, Kreutz EW, Michel J, Niessen M, Kostrykin V, Schulz W, Etzkorn A, Bobzin K, Lugscheider E, Poprawe R (2000) Microholes in Zirconia Coated Ni-Superalloys for Transpiration Cooling of Turbine Blades. Proceedings of the SPIE 4065, High-Power Laser Ablation III, 218.

88. Hrabe N, Quinn T (2013) Effects of Processing on Microstructure and Mechanical Properties of a Titanium Alloy (Ti6Al4V) Fabricated Using EBM, Part 1: DISTANCE from Build Plate and Part Size. Materials Science and Engineering: A 573:264–270.

89. Hrabe N, Quinn T (2013) Effects of Processing on Microstructure and Mechanical Properties of a Titanium Alloy (Ti–6Al–4V) Fabricated Using Electron Beam Melting (EBM). Part 2: Energy Input, Orientation, and Location. Materials Science and Engineering: A 573:271–277.

90. Hubscher B (2012) NASA's Space Launch System Using Futuristic Technology to Build the Next Generation of Rockets.

91. Innovative Technologies for Future Alloys (2013). Retrieved from http:// www.turbinentechnik.com/files/alloy_folder.pdf.

92. Izquierdo B, Plaza S, Sanchez JA, Pombo I, Ortega N (2012) Numerical Prediction of Heat Affected Layer in the EDM of Aeronautical Alloys. Applied Surface Science 259:780–790.

93. Jabbaripoura B, Sadeghia MH, Shabgardb MR, Faraji H (2013) Investigating Surface Roughness Material Removal Rate and Corrosion Resistance in PMEDM of-TiAl Intermetallic. Journal of Manufacturing Processes 15: 56–68.

94. Jain VK, Chavan A, Kulkarni A (2007) Experimental and Analytical Study of Contoured Holes by Shaped Tube Electrochemical Drilling

Process. Proceedings of the 15th Int Symp on Electromachining, Pittsburgh, USA, 315–318. 724 F. Klocke et al. / CIRP Annals - Manufacturing Technology 63 (2014) 703–726

95. Jawahir IS, Brinksmeier E, M'Saoubi R, Aspinwall DK, Outerio JC, Meyer D, Umbrello D, Jayal AD (2011) Surface Integrity in Material Removal Processes: Recent Advances. CIRP Annals – Manufacturing Technology 60(2):603–626.

96. Jefferies R (2013) Continuous Lower Energy, Emissions and Noise (CLEEN) Program. USACA Spring Association Meeting, 21 May.

97. Jones J, McNutt P, Tosi R, Perry C, Wimpenny D (2012) Remanufacture of Turbine Blades by Laser Cladding, Machining and In-Process Scanning in a Single Machine. Retrieved from www.dora.dmu.ac.uk.

98. Joyce D (2012) GE Aviation. Barclays Capital Industrial Select Conference.

99. Kao JY, Tsao CC, Wang SS, Hsu CY (2010) Optimisation of the EDM Parameters on Machining Ti–6Al–4V with Multiple Quality Characteristics. International Journal of Advanced Manufacturing Technology 47:395–402.

100. Kelbassa I, Albus P, Dietrich J, Wilkes J (2008) Manufacture and Repair of Aero Engine Components Using Laser Technology. Proceedings of the 3rd Pacific Conference on Application of Lasers and Optics, 208–212.

101. Klocke F, König W (2007) Fertigungsverfahren 3: Abtragen, Generieren und Lasermaterialbearbeitung, Springer, Berlin. ISBN 3-540-23492-6.

102. Klocke F, Welling D, Dieckmann J (2011) Comparison of Grinding and WireEDM Concerning Fatigue Strength and Surface Integrity of Machined Ti6Al4V Components. Procedia Engineering 19:184–189.

103. Klocke F, Welling D, Dieckmann J, Veselovac D, Perez R (2012) Developments in Wire-EDM for the Manufacturing of Fir Tree Slots in Turbine Discs Made of Inconel 718. Key Engineering Materials 504–506:1177–1182.

104. Klocke F, Welling D, Klink A, Veselovac D, Nöthe T, Perez R (2014) Evaluation of Advanced Wire-EDM Capabilities for the Manufacture of Fir Tree Slots in Inconel 718. Submitted to 6th

CIRP International Conference on High Performance Cutting, Berkely, USA.

105. Klocke F, Zeis M, Harst S, Herrig T, Klink A (2013) Analysis of the Simulation Accuracy of Electrochemical Machining Processes Based on the Integration Level of Different Physical Effects. M. Scripts Precision and Microproduction Engineering 7, Fraunhofer IWU, 165–170. ISBN 978-3-942267-95-3.

106. Klocke F, Zeis M, Harst S, Klink A, Veselovac D, Baumga¨rtner M (2013) Modeling and Simulation of the Electrochemical Machining (ECM) Material Removal Process for the Manufacture of Aero Engine Components. Procedia 8:265–270.

107. Klocke F, Zeis M, Klink A (2012) Technological and Economical Capabilities of Manufacturing Titanium- and Nickel-Based Alloys Via Electrochemical Machining (ECM). Key Engineering Materials 504–506:1237–1242.

108. Klocke F, Zeis M, Klink A, Veselovac D (2012) Technological and Economical Comparison of Roughing Strategies via Milling, EDM and ECM for Titaniumand Nickel-Based Blisks. Procedia CIRP 2:98–101.

109. Klocke F, Zeis M, Klink A, Veselovac D (2013) Experimental Research on the Electrochemical Machining of Modern Titanium- and Nickel-Based Alloys for Aero Engine Components. Procedia CIRP 6:369–373.

110. Klocke F, Zeis M, Klink A, Veselovac D (2013) Technological and Economical Comparison of Roughing Strategies Via Milling, Sinking-EDM, Wire-EDM and ECM for Titanium- and Nickel-Based Blisks. CIRP JMST 6(3):198–203.

111. Krauss H, Eschey C, Za¨h MF (2012) Thermography for Monitoring the Selective Laser Melting Process. Solid Freeform Fabrication: ProceedingsUniversity of Texas, Austin.

112. Krauss H, Za¨h MF (2013) Multi-target Optimization and Process Window Analysis in Selective Laser Melting of High-performance Parts. 22nd International Conference on Production Research (ICPR 22).

113. Kremer D, Bazine G, Moisan A, Bessaguet L, Astier A, Thanh NK (1983) Ultrasonic Machining Improves EDM Technology. Proceedings of the 7th International Symposium on

Electromachining (ISEM VII), 2–14 April, Birmingham, UK, 67–76.

114. Kremer D, Lhiaubet C, Moisan A (1991) A Study of the Effect of Synchronizing Ultrasonic Vibrations with Pulses in EDM. Annals of CIRP 40(1):211–214.

115. Kruth JP, Levy G,Klocke F, Childs THC (2007) ConsolidationPhenomena in Laser and Powder-Bed Based Layered Manufacturing. CIRP Annals 56(2):730–759.

116. Kunieda M, Lauwers B, Rajurkar KP, Schumacher BM (2005) Advancing EDM Through Fundamental Insight into the Process. Annals of CIRP 54(2):64–87.

117. Kuriakose S, Shunmugam MS (2004) Characteristics of Wire-Electro Discharge Machined Ti6Al4V Surface. Materials Letters 58:2231–2237.

118. Lakomiec M (2013) Serienfertigung von Triebwerksteilen mittels Laserstrahlschmelzen, VDI Fachkonferenz Additive Manufacturing – Vom Prototypen bis zur Großserie mit generativen Fertigungsverfahren, Duisburg.

119. Lauwers B, Klocke F, Klink A, Tekkaya AE, Neugebauer R, Mcintosh D (2014) Hybrid Processes in Manufacturing. CIRP Annals – Manufacturing Technology 63(2):561–583.

120. Leahy J (2013) Global Market Forecast 2013–2032. www.airbus-group.com.

121. Lee HG, Simao J, Aspinwall DK, Dewes RC, Voice W (2004) Electrical Discharge Surface Alloying. Journal of Materials Processing Technology 149:334–340.

122. Leigh S, Sezer K, Li L, Grafton-Reed C, Cuttell M (2010) Recast and Oxide Formation in Laser-Drilled Acute Holes in CMSX-4 Nickel Single-Crystal Superally. Proceedings of the IMechE, Part B: Journal of Engineering Manufacture 224 7:1005–1016.

123. Levy GN (2010) The Role and Future of the Laser Technology in the Additive Manufacturing Environment. Physics Procedia 5:65–80.

124. Li J, Wang HM (2010) Microstructure and Mechanical Properties of Rapid Directionally Solidified Ni-Base Superalloy Rene 41 by Laser Melting Deposition Manufacturing. Materials Science and Engineering: A 527:4823–4829.

125. Li L, Guo YB, Wei XT, Li W (2013) Surface Integrity Characteristics in WireEDM of Inconel 718 at Different Discharge Energy. Procedia CIRP 6:221–226.

126. Lin YC, Yan BH, Chang YS (2000) Machining Characteristics of Titanium Alloy (Ti–6Al–4V) Using a Combination Process of EDM with USM. Journal of Materials Processing Technology 104:171–177.

127. Liu K, Ferraris E, Peirs J, Lauwers B, Reynaerts D (2008) Precision Manufacturing of the Ultra Miniature Gas Turbine in Ceramic Composite for the Micro Power Generation System. Proceedings of the Euspen International Conference, Zurich.

128. Liu X, Kang X, Zhao W, Liang W (2013) Electrode Feeding Path Searching for 5-Axis EDM of Integral Shrouded Blisks. Procedia CIRP 6:107–111.

129. 132] Majundar JD (2012) Laser Assisted-Fabrication of Materials, Springer, Berlin.

130. 133] Mantle AL, Abboud E, Aspinwall DK (1997) Productivity and Workpiece Surface Integrity Effects When Electrical Discharge Wire Machining a Gamma Titanium Aluminide. Proceedings of the 14th Conference of the Irish Manufacturing Committee (IMC-14), 3–5 September, Trinity College Dublin, Ireland, 443–450.

131. Meiners W (1999) Direktes Selektives Laser-Sintern Einkomponentiger Metallischer Werkstoffe, Shaker, Aachen.

132. Meiners W (2012) High Power SLM Machining of Inconel 718, Annual Report, Fraunhofer-Institut für Lasertechnik ILT.

133. Metcut Research Associates (1980) Machining Data Handbook, 3rd ed., vol. 2. MDC, Cincinnati, OH.

134. Mühlbauer G (2011) Funkenerosion im Triebwerksbau. Workshop Mikrofunkenerosion 14, TU, Berlin. April.

135. Mumtaz K, Hopkinson N (2010) Selective LaserMeltingofThinWallParts Using Pulse Shaping. Journal of Materials Processing Technology 210(2):279–287.

136. Mumtaz KA, Hopkinson N (2007) Laser Melting Functionally Graded Composition of Waspaloy and Zirconia Powders. Journal of Material Science 42(18):7647–7656.

137. Murr LE, Gaytan SM, Ceylan A, Martinez E, Martinez JL, Hernandez DH, Machado BI, Ramirez DA, Medina F, Collins

S, Wicker RB (2010) Characterization of Titanium Alluminide Alloy Components Fabricated by Additive Manufacturing Using Electron Beam Melting. Acta Materialia 58:1887–1894.

138. Murthy VSR, Philip PK (1983) Pulse Train Analysis in Ultrasonic Assisted EDM. International Journal of Machine Tools and Manufacture 27(4):469–477.

139. NCMT (2013) Deep-Hole EDM Drilling of Turbine Components is Seven Times Faster. Retrieved from http://www.ncmt.co.uk.

140. Okane M, Goto A (2008) Development of MS Coating for Aircraft Engine Parts. Mitsubishi Electric Advance Mechatronics 123. ISSN 1345-3041.

141. Okasha MM, Mativenga PT, Driver N, Li L (2010) Sequential Laser and Mechanical Micro-Drilling of Ni Superalloy for Aerospace Application. CIRP Annals – Manufacturing Technology 59:199–202.

142. Ott M (2012) Multimaterialverarbeitung bei der additiven strahl- und pulverbettbasierten Fertigung, Dissertation, TU, Mu''nchen.

143. Ott M, Zaeh MF (2010) Multi-Material Processing in Additive Manufacturing. in Bourell DL, et al.(Eds.) 2010 – Proceedings of the 21st SFF Symposium, Austin, University of Texas at Austin 195–203.

144. Pandey A, Singh S (2010) Current Research Trends in Variants of Electrical Discharge Machining: A Review. International Journal of Engineering Science and Technology 2(6):2172–2191.

145. Pattavanitch J, Hinduja S (2012) Machining of Turbulated Cooling Channel Holes in Turbine Blades. CIRP Annals – Manufacturing Technology 61:199–202.

146. Paul MA, Aspinwall DK (1998) Arc Sawing Performance Evaluation and Machine Design. Proceedings of the 12th International Symposium on Electromachining (ISEM XII), 11–13 May, Aachen, Germany, 407–416.

147. Paul MA, Liang WS, Aspinwall DK (1998) Arc Sawing Applications and the Development of Arc Profile Machining. Proceedings of the 4th International Conference on Advances in Materials and Processing Technology (AMPT), 24–28 August, Kuala Lumpur, Malaysia, 819–826.

148. Platz A, Feiling N (2013) Precise Electrochemical Machining of Nickel Base Integrated Blade Compressor Rotors, Precision and Microproduction Engineering 7, Fraunhofer IWU, Chemnitz23–32. ISBN 978-3-942267-95-3.

149. Poprawe R (2005) Lasertechnik fu¨r die Fertigung-Grundlagen, Perspektiven und Beispiele fu¨r den innovativen Ingenieur, Springer, Berlin.

150. Poprawe R, Kelbassa I, Walther K, Witty M, Bohn D, Krewinkel R (2008) Optimising and Manufacturing a Laser-Drilled Cooling Hole Geometry for Effusion-Cooled Multi-Layer Plates. Proceedings of the 12th ISROMAC Hawai, 17–22 February, 1–10.

151. Prihandana GS, Mahardika M, Hamdi M, Mitsui K (2011) Effect of LowFrequency Vibration on Workpiece in EDM Processes. Journal of Mechanical Science and Technology 25(5):1231–1234.

152. Qi H, Azer M, Singh P (2010) Adaptive Toolpath Deposition Method for Laser Net Shape Manufacturing and Repair of Turbine Compressor Airfoils. International Journal of Advanced Manufacturing Technology 48:121–131.

153. Ramakrishnan R, Karunamoorthy L (2008) Modeling and Multi-Response Optimization of Inconel 718 on Machining of CNC WEDM Process. Journal of Materials Processing Technology 207:343–349.

154. Reed RC (2006) The Superalloys, Cambridge University Press, Cambridge.

155. Richter KH (2008) Laser Material Processing in the Aero Engine Industry. Established Cutting-Edge and Emerging Applications. Proceedings of the 3rd Pacific International Conference on Application of Lasers and Optics.

156. Richter KH (2010) Using the Laser for Build-Up. Established and Emerging Applications at MTU Aero Engines. International Laser Technology Congress AKL'10, Aachen.

157. Richter KH, Orban S, Nowotny S (2004) Laser Cladding of the Titanium Alloy Ti6242 to Restore Damaged Blades. Proceedings of the 23rd International Congress on Applications of Lasers and Electro-Optics.

158. Rodriguez E, Medina F, Espalin D, Terrazas C, Muse D, Henry C, MacDonald E, Wicker R (2012) Integration of a Thermal Imaging

Feedback Control System in Electron Beam Melting. Solid Freeform Fabrication: ProceedingsUniversity of Texas, Austin.

159. Rolls Royce – The Jet Engine, (2005), . ISBN: 0902121235.

160. Rolls Royce Market Outlook 2012-31 (2012) www.rolls-royce.com.

161. Sánchez JA, Plaza S, Gil R, Ramos JM, Izquierdo B, Ortega N, Pombo I (2013) Electrode Set-Up for EDM-Drilling of Large Aspect-Ratio Microholes. Procedia CIRP 6:275–280.

162. Sarkar S, Mitra S, Bhattarcharyya B (2005) Parametric Analysis and Optimization of Wire Electrical Discharge Machining of g-Titanium Aluminide Alloy. Journal of Materials Processing Technology 159:286–294.

163. Sarkar S, Sekh M, Mitra S, Bhattacharyya B (2008) Modeling and Optimization of Wire Electrical Discharge Machining of gTiAl in Trim Cutting Operation. Journal of Materials Processing Technology 205:376–387.

164. Schubert A, Hackert-Oschätzchen M, Meichsner G, Zinecker M, Edelmann J (2011) Precision and Micro ECM with Localized Anodic Dissolution. in Slabe JM, (Ed.) TECOS Slovenian Tool and Die Development Centre, Celje: Proceedings of the 8th International Conference on Industrial Tools and Material Processing Technologies, 193–196. ISBN: 978-961-6692-02-1.

165. Schwerdtfeger JV, Singer RF, Körner C (2012) In Situ Flaw Detection by IRImaging During Electron Beam Melting. Rapid Prototyping Journal 18(4):259– 263.

166. Sen B, Kiyawat N, Singh PK, Mitra S, Ye JH, Purkait P (2003) Developments in Electric Power Supply Configurations for Electrical-Discharge-Machining (EDM). Proceedings of the Fifth International Conference on Power Electronics and Drive Systems (PEDS 2003) – Vol. 1, 17–20 November, Singapore, 659–664.

167. Sen M, Shan HS (2005) A Review of Electrochemical Macro- to Micro-Hole Drilling Processes. International Journal of Machine Tools & Manufacture 45:137–152.

168. Sezer HK, Li L, Leigh S (2009) Twin Gas Jet-Assisted Laser Drilling Through Thermal Barrier-Coated Nickel Alloy Substrates. International Journal of Machine Tools & Manufacture 49:1126–1135.

169. 173. Shen N, Chou K (2012) Numerical Thermal Analysis in Electron Beam Additive Manufacturing with Preheating Effects. Solid Freeform Fabrication: ProceedingsUniversity of Texas, Austin.

170. Shepeleva L, Medres B, Kaplan WD, Bamberger M, Weisheit A (2000) Laser Cladding of Turbine Blades. Surface and Coatings Technology 125:45–48.

171. Sieber J (2014) Langfristige Sicherung des Luftverkehrs durch neue Antriebstechnologien und alternative Brennstoffe. Retrieved from www.mtu.de.

172. Sivakumar KM, Gandhinathan R (2013) Establishing Optimum Process Parameters for Machining Titanium Alloys (Ti–6Al–4V) in Spark Electrical Discharge Machining. International Journal of Engineering and Advanced Technology 2(4):201–204.

173. Soo SL, Antar MT, Aspinwall DK, Sage C, Cuttell M, Perez R, Winn AJ (2013) The Effect of Wire Electrical Discharge Machining on the Fatigue Life of Ti– 6Al–2Sn–4Zr–6Mo Aerospace Alloy. Procedia CIRP 6:215–219.

174. Soo SL, Hood R, Aspinwall DK, Voice WE, Sage C (2011) Machinability and Surface Integrity of RR1000 Nickel Based Superalloy. CIRP Annals – Manufacturing Technology 60:89–92.

175. Steffens K, Platz A, Buckl F, (2004) Feinbearbeitungsverfahren – Schlu¨ sseltechnologien fu¨ r moderne Luftfahrtverdichter. MTU Aero Engines. www.mtu.de.

176. Steffens K, Walther R (2003) Driving the Technological Edge in Airbreathing Propulsion. Retrieved from www.mtu.de.

177. Steffens K, Wilhelm H (2013) Werkstoffe, Oberfla¨chentechnik und Fertigungsverfahren fu¨r die na¨chste Generation von Flugtriebwerken. Retrieved from www.mtu.de.

178. Stimper B (2014) Using Laser Powder Cladding to Build Up Worn Compressor Blade Tips MTU Aero Engines. Retrieved from www.mtu.de.

179. Su W (2002) Layered Fabrication of Tool Steel and Functionally Graded Materials with a Nd: YAG Pulsed Laser, (PhD thesis) Loughborough University, Loughborough, UK.

180. Taberno I, Lamikiz A, Martinez S, Ukar E, Figueras J (2011) Evaluation of the Mechanical Properties of Inconel 718

Components Built by Laser Cladding. International Journal of Machine Tools & Manufacture 51:465–470.

181. Thanigaivelan R, Arunachalam RM, Karthikeyan B, Loganathan P (2013) Electrochemical Micromachining of Stainless Steel with Acidified Sodium Nitrate Electrolyte. Procedia CIRP 6:352–356.

182. Thoe TB, Aspinwall DK, Killey N (1999) Combined Ultrasonic and Electrical Discharge Machining of Ceramic Coated Nickel Alloy. Journal of Materials Processing Technology 92–93:323–328.

183. Thümmler T (2008) Herstellung von komplexen Kühlluftbohrungen in Hochdruckturbinenschaufeln, MTU Aero Engines.

184. Titan beim Zerspanen nicht länger verschwenden. Produktion 41:18. www.produktion.de.

185. 189. Toller DF (1983) Multi-Small Hole Drilling by EDM. Proceedings of the 7th International Symposium on Electromachining (ISEM VII), 12–14 April, Birmingham, UK, 147–156.

186. Toyserkani E, Khajepour A (2005) Laser Cladding, CRC Press, Boca Raton.

187. Uhlmann E, Domingos DC (2013) Development and Optimization of the DieSinking EDM-Technology for Machining the Nickel-Based Alloy MAR-M247 for Turbine Components. Procedia CIRP 6:181–186.

188. Uhlmann E, Domingos DC (2013) Investigations on Vibration-Assisted EDM – Machining of Seal Slots in High-Temperature Resistant Materials for Turbine Components. Procedia CIRP 6:71–76.

189. Uhlmann E, Oberschmidt D, Langmack M (2013) Complex Bore Holes fabricated by combined Helical Laser Drilling and Micro Electrical Discharge Machining. 28th ASPE Annual Meeting.

190. van Tijum R, Pajak T (2008) The Multiphysics Approach: The Electrochemical Machining Process, Presentation During COMSOL Conference Hannover.

191. Verfahren zur Reduzierung von Chrom(VI) bei der ECM-Bearbeitung, Retrieved from http://www.maschinenmarkt.vogel.de

192. Veselovac D (2013) Process and Product Monitoring in the Drilling of Critical Aero Engine Components, (Dissertation) WZL RWTH Aachen University.

193. Vrancken B, Thijs L, Kruth J-P, van Humbeeck J (2012) Heat Treatment of Ti6Al4V Produced by Selective Laser Melting: Microstructure and Mechanical Properties. Journal of Alloys and Compounds 541:177–185.

194. Walther R (2012) Recent Challenges in Air Breathing Propulsion. Proceedings of the 14th International Symposium on Transport Phenomena and Dynamics of Rotating Machinery.

195. Wang JY, Yu Y, McGeough JA, De Silva A (2007) Experimental Investigation for the Enhancement of Accuracy of Pulse Electrochemical Machining by Improvement of Pulse Power. Proceedings of the 15th ISEM, Pittsburgh, USA, 369–373.

196. Wang MH, Zhub D (2009) Simulation of Fabrication for Gas Turbine Blade Turbulated Cooling Hole in ECM Based on FEM. Journal of Materials Processing Technology 209:1747–1751.

197. Wang Z, Guan K, Gao M, Li X, Chen X, Zeng X (2012) The Microstructure and Mechanical Properties of Deposited-IN718 by Selective Laser Melting. Journal of Alloys and Compounds 513:518–523.

198. Wardono B, Ismail MFB, Liew PJ (2011) An Analysis of EDM Die Sinking Parameters on Ti–6Al–4V. Proceedings of the International Conference and Exhibition on Sustainable Energy and Advanced Materials (ICESEAM2011), Solo, Indonesia, 3–4 October, 367–373.

199. Wei B, Kozak J, Rajurkar KP (1994) Pulse Electrochemical Machining (PECM) of Ti–6Al–4V Alloy. Transactions of NAMRI/SME XXII 141–147.

200. Wei B, Trimmer AL, Luo Y, Yuan R, Hayashi S, Lamphere M (2010) Advancement in High Speed Electro-Erosion Processes for Machining Tough Metals. Proceedings of the 16th ISEM, 19–23 April, Shanghai, China, 193–196.

201. Welling D (2013) Drahtfunkenerosive Bearbeitung von Profilnuten in Nickelbasislegierungen. 9. Fachtagung Funkenerosion, RWTH Aachen University.

202. Welling D (2014) Results of Surface Integrity and Fatigue Study of Wire-EDM compared to Broaching and Grinding for demanding Jet Engine Components Made of Inconel 718. Procedia CIRP 13:339–344.

203. Westley JA, Atkinson J, Duffield A (2004) Generic Aspects of Tool Design for Electrochemical Machining. Journal of Materials Processing Technology 149:384–392.

204. Winbro Group Technologies (2014) Series 800 Laser & EDM Datasheet.

205. Witzel J, Schopphoven T, Gasser A, Kelbassa I (2011) Development of a Model for Prediction of Material Properties of Laser Cladded Inconel 718 as Related to Porosity in the Bulk Material. Proceedings of the 30th ICALEO, Orlando, USA, 275–282.

206. Witzel J, Schrage J, Gasser A, Kelbassa I (2011) Additive Manufacturing of a Blade-Integrated Disk by Laser Metal Deposition. Proceedings of 30th International Congress on Applications of Lasers & Electro-Optics, Orlando, USA, 250–256.

207. Wohlers T, Gornet T (2011) History of Additive Manufacturing. in Wohlers T, (Ed.)Wohlers Report: Additive Manufacturing and 3DPrinting State ofthe Industry Annual Worldwide Progress Report, Wohlers Associates, Fort Collins, CO.

208. Wohlers TT (2012) Wohlers Report 2012: Additive Manufacturing and 3D Printing State of the Industry: Annual Worldwide Progress Report, Wohlers Associates, Fort Collins, CO.

209. Wollenberg G, Schulze HP, Trautmann HJ, Kappmeyer G (2007) Controlled Current Rise for Pulsed Electrochemical Machining. Proceedings of the 15th International Symposium on Electromachining, Pittsburgh, USA.

210. Xu ZY, Xu Q, Zhu D, Gong T (2013) A High Efficiency Electrochemical Machining Method of Blisk Channels. CIRP Annals – Manufacturing Technology 62:187–190.

211. Xue L, Li Y, Wang S (2011) Direct Manufacturing of Net-Shape Functional Components/Test-Pieces for Aerospace, Automotive and Other Applications. Proceedings of the 30th ICALEO, 23–27 October, Orlando, USA, 479–488.

212. Yang DY, Cao FG, Liu YJ, Yang LG, Zhang K, Zhu YF (2013) Overview on FiveAxis Precision EDM Techniques. Procedia CIRP 6:193–199.

213. Yasa E, Kruth JP, Deckers J (2011) Manufacturing by Combining Selective Laser Melting and Selective Laser Erosion/Laser Re-melting. CIRP Annals 60:263–266.

214. Ye L (1992) Apparatus for electrical machining. US Patent 5128010/1992.

215. Yu J, Rombouts M, Motmans F (2012) Material Properties of Ti6Al4V Parts Produced by Laser Metal Deposition. Physics Procedia 39:416–424.

216. Yu J, Xiao P, Liao Y, Cheng M (2009) Surface Integrity in Electrical Discharge Machining of Ti–6Al–4V. Advanced Materials Research 76–78:613–617.

217. Yuan R, Wei B, Luo Y, Zhan Y, Xu W, Lamphere M (2010) Advancement in High Speed Electro-Erosion Processes for Machining Tough Metals. Proceedings of the 16th International Sympsium on Electromachining (ISEM), 19–23 April, Shanghai, China, 207–210.

218. Zaeh MF, Branner G (2010) Investigations on Residual Stresses and Deformations in Selective Laser Melting. Production Engineering – Research and Develepment 4(1):35–45.

219. Za¨h MF, Lutzmann S (2010) Modelling and Simulation of Electron Beam Melting. Production Engineering – Research and Develepment 4(1):15–23.

220. Za¨h MF, Ott M (2011) Investigations on Heat Regulation of Additive Manufacturing Processes for Metal Structures. CIRP Annals – Manufacturing Technology 60(1):259–262.

221. Zhao JS, Xu JW, Zhu YW (2007) Design Optimization of Cathode's Feeding Path of NC-Electrochemical Machining Based on Computer Simulation of Shaping Process. Proceedings of the 15th International Symposium on Electromach, Pittsburgh, USA, 365–368.

222. Zhao W, Go L, Xu H, Li L, Xiang X (2013) A Novel High Efficiency Electrical Erosion Process – Blasting Erosion Arc Machining. Procedia CIRP 6:622–626.

223. Zheng Z, Wang Y, Dong Y, Wang Z (2010) Fabrication of Key Components in Micro Turbine Engine by Using Micro Electrical Discharge Machining (EDM). Proceedings of the 16th International Symposium on Electromachining, Shanghai, 587–591.

224. Zhu D, Zhu D, Xu Z, Zhou L (2013) Trajectory Control Strategy of Cathodes in Blisk Electrochemical Machining. Canadian Journal of Anaesthesia 26(4):1064–1070.

225. Hofmann DC, Borgonia JP, Dillon RP, Suh EJ, Mulder JL, Gardner PB (2013) Methods for Fabricating Gradient Alloy Articles with Multi-Functional Properties, Patent WO 2013/112217 A2.

Experimental Study of a Mixed Refrigerant Joule–Thomson Cryocooler using a Commercial Air-Conditioning Scroll Compressor

Jisung Lee[a, 1], Kyungsoo Lee[a, b], and Sangkwon Jeong[a, 2]

[a]Cryogenic Engineering Laboratory, Division of Mechanical Engineering, School of Mechanical, Aerospace and Systems Engineering, Korea Advanced Institute of Science and Technology, 335 Gwahangno, Yuseong-gu, Daejeon 305-701, Republic of Korea

[b]Department of Internal Medicine, University of Michigan, Ann Arbor, 4520 MSRB I, 1150 W. Medical Center Dr., Ann Arbor, MI 48109, USA

ABSTRACT

Mixed refrigerant Joule–Thomson (MR J–T) cryocoolers have been used to create cryogenic temperatures and are simple, efficient, cheap, and durable. However, compressors for MR J–T cryocoolers still require optimization. As the MR J–T cryocooler uses a commercial scroll compressor developed for air-conditioning systems, compressor overheating due to the use of less optimized refrigerants may not be negligible, and could cause compressor malfunction due to burn-out of scroll tip seals. Therefore, in the present study, the authors propose procedures to optimize compressor operation to avoid the overheating issue when the MR J–T cryocooler is used with a commercial oil lubricated scroll compressor, and the present experimental results obtained for a MR J–T cryocooler. A single stage 1.49 kW (2 HP) scroll compressor designed for R22 utilizing a mixture of nitrogen and hydrocarbons was used in the present study. As was expected, compressor overheating and irreversible high temperatures at a compressor discharge port were found at the beginning of compressor operation, which is critical, and hence, the authors used a water injection cooling system for the compressor to alleviate temperature overshooting. In addition, a portion of refrigerant in the high-pressure stream was by-passed into the compressor suction port. This allowed an adequate compression ratio, prevented excessive temperature increases at the compressor discharge, and eventually enabled the MR J–T cryocooler to operate stably at 121 K. The study shows that commercial oil lubricated scroll compressors can be used for MR J–T cryocooling systems if care is exercised to avoid compressor overheating.

INTRODUCTION

When high-pressure fluid is expanded adiabatically, such as, through a narrow orifice, the fluid is expanded isenthalpically and fluid temperature is changed referred to as the Joule–Thomson (J–T) effect. For fluids other than helium, neon, and hydrogen, temperature decreases when the fluid is expanded from ambient temperature. J–T cryocoolers that use this Joule–Thomson effect have advantages such as simple designs and no moving components in the cold region, high reliability, low vibration, and low manufacturing cost [1]. However, the high compression ratios required to achieve an adequate cooling effect are considered a primary disadvantage of J–T cryocoolers, and thus, efforts have been made to find appropriate ways of decreasing compression ratios and increasing cooling effect by using mixed refrigerant (MR), whereby high boiling point refrigerants are added to low boiling point refrigerants. The operation of J–T cryocoolers with MRs have been reported to increase system efficiency, and studies on these cryocoolers aimed at optimizing the composition and cycle configuration for specific cooling purposes are still interesting topics of research [2], [3] and [4].

It has been reported that the efficient operation of MR J–T cryocoolers at a high pressure of 2500 kPa is an important consideration for popular applications, because this pressure range can be easily achieved using a commercial oil lubricated compressor. There are mainly three types of compressors, such as, reciprocating, rotary, and scroll for air-conditioning applications. Scroll compressors are known as low vibration and low noise by its unique operating mechanism of continuous compression process and no valves in suction and discharge port. Although commercial oil lubricated rotary compressors have been successfully used for MR J–T cryocoolers [5], [6], [7] and [8], scroll compressors were found to be irreversibly damaged during preliminary experiments due to compressor overheating. Thus, the authors modified a MR J–T unit to reduce this overheating by adding a by-pass line connecting the compressor suction and discharge ports. Here, we describe the technical aspects and the principles underlying the operation of the modified MR J–T unit. In addition, experimental results regarding MR J–T operation are presented.

DESCRIPTION OF EXPERIMENTAL APPARATUS AND METHOD

Experimental Circuit

The experimental refrigeration circuit is depicted in Fig. 1, and includes a scroll compressor, oil separators and oil mist filters, a heat exchanger, and a J–T expansion valve. The scroll compressor (ZR23K1-PFV, Copeland, USA) used in the present study had a capacity of 1.49 kW (2 HP). ZEROL 150 oil (Shrieve Chemical Products, USA) was used for compressor lubrication. The carbonization temperature of this oil is above 448 K, and its freezing point is 203 K, and thus it is recommended for compressors operating with hydrocarbon and ammonia gases. The lubrication oil contained in the refrigerant must be separated out to avoid clogging issues at the coldest part of the cryocooler. Thus, two oil separators (OUB1, Danfoss, Japan) were installed in sequence prior to the aftercooler, because the oil is separated more effectively at higher temperatures, and thus, lower viscosities. A portion of refrigerant was assumed to be condensed at stream 4 (Fig. 1), and remaining oil mists contained in the refrigerant were removed using the oil mist filters. The oil mist filters were composed of two different kinds of glass microfiber filters (Whatman, Maidstone, UK), that is, GF/C 1.2 μm filters for the first two stages and a GF/F 0.7 μm filter for the final stage (Fig. 2).

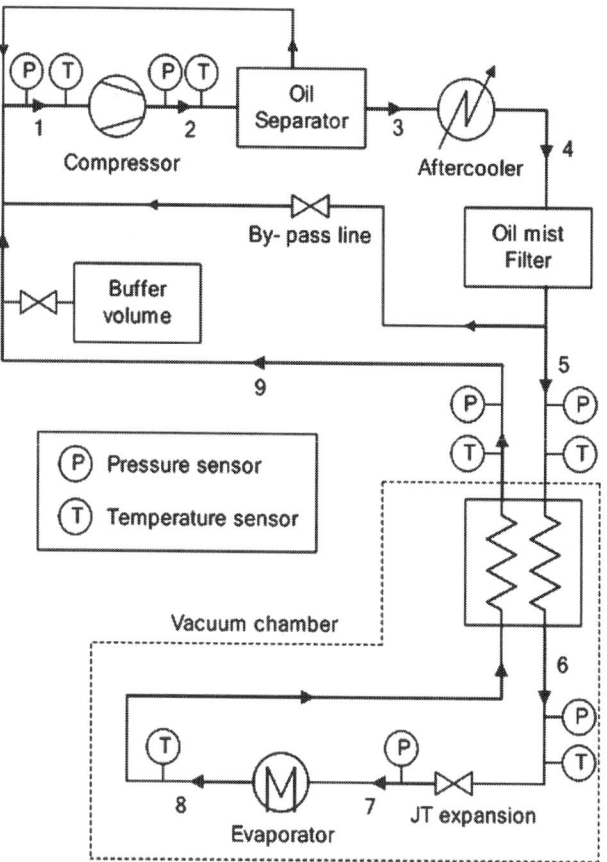

Figure 1: Cycle configuration of MR J–T cryocooler.

Two O-ring grooves

Filter holder

Figure 2: Structure and configuration of domestically fabricated oil filter system.

High pressure refrigerants are precooled by the low temperature returning stream through a tubes-in-tube counter flow heat exchanger fabricated from 1/4 to 3/8 in. copper tubes. Total length of the heat exchanger was approximately 12 m, and a spacer was placed between the inner and outer tubes to maximize heat exchange. High pressure refrigerant flows through the outer lumen of the heat exchanger, whereas the low pressure stream flows through the inner lumen. This arrangement allowed a minimal pressure drop between inlets and outlets of the heat exchanger. In addition, a bypass pathway was integrated into the stream 5 (Fig. 1) to shunt a portion of the high pressure stream refrigerant into the low pressure stream under valve control. The mixed refrigerant at the exit of the heat exchanger is then expanded isenthalpically through the expansion valve. The expanded low temperature refrigerants absorb heat through the evaporator, where a manganin wire heater (MW-36, LakeShore, USA) was installed as a heating load. After the evaporator, the low pressure stream returns to the compressor at ambient temperature. The cryogenic temperature parts including the heat exchanger, the J–T expansion unit, and the evaporator were covered by several layers of radiation shield and sealed in a vacuum chamber to minimize heat invasion.

In addition, a 4.5 L refrigerant reservoir was integrated at the suction line of the compressor. This was introduced with the aim of increasing refrigerant density in the circuit when the refrigerant is condensed. Refrigerants were charged into the reservoir at an initial charging pressure of 1150 kPa, and were replenished into the principal circuit when the circuit pressure decreased. Temperature and pressures at each point were measured using K-type thermocouples and commercial pressure transducers (Sensym ST2500G1, Honeywell, USA). The accuracies of these temperature and pressure sensors were ±1 K and ±17.2 kPa (2.5 psi), respectively.

Mixed Refrigerant and Its Composition

Experiments were performed using a gas mixture of nitrogen and four different hydrocarbon refrigerants, namely, methane, ethane, propane, and iso-butane. The gas mixture had percentage molar ratios of 15%, 30%, 30%, 10%, and 15% for nitrogen, methane, ethane, propane, and iso-butane, respectively. Enthalpies of the refrigerant mixture at 2000 and 300 kPa were calculated with respect to temperature, and the isothermal enthalpy difference is presented in Fig. 3. The enthalpy difference increased and peaked in an intermediate temperature around 250 K, but cooling capacity of MR-JT cryocooler was limited to the minimum enthalpy difference over the whole temperature range [9]. Therefore, the plot of enthalpy difference can be used to estimate cooling capacity as the below of the dashed line. For ideal operation, the gas mixture was assumed to achieve a minimum temperature of 94 K and to have a cooling capacity of 8.5 J/g at 100 K. The required compositions of refrigerants were charged into the refrigeration circuit by partial pressure ratio, at a target total charging pressure of 1150 kPa. In addition, a portion of mixed refrigerant was withdrawn immediately after MR J–T experiments, and compositions were determined by gas chromatography (HP 5890 series II, USA) using a packed type Porapak-Q column. Helium was used as the carrier gas at a flow rate of 30 cc/min, and the oven temperature was programmed from 80 °C to 150 °C over 10 min. A flame ionization detector (FID) was used, and two types of known compositions standard gases were analyzed before analyzing the sample gas compositions.

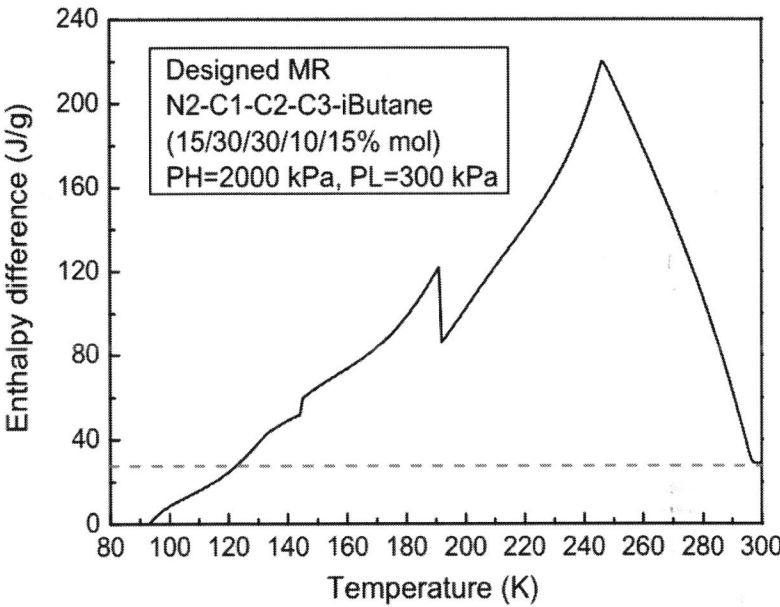

Figure 3: Enthalpy difference between high pressure and low pressure with selected mixture [11].

EXPERIMENTAL RESULT

The pressure profiles of the MR J–T cryocooler are described in Fig. 4; these pressure profiles were measured without refrigerant by-pass. System pressures gradually increased during the first 150 s of compressor operation. Subsequently, the J–T expansion valve was regulated to increase the compression ratio, and at the same time, the compressor exit temperature rapidly increased to over 400 K. Compressor power was turned off to protect the compressor from excessive heat, and the compressor was restarted when the discharge temperature decreased to ambient temperature. However, the obtained compression ratio during the re-operation was negligible.

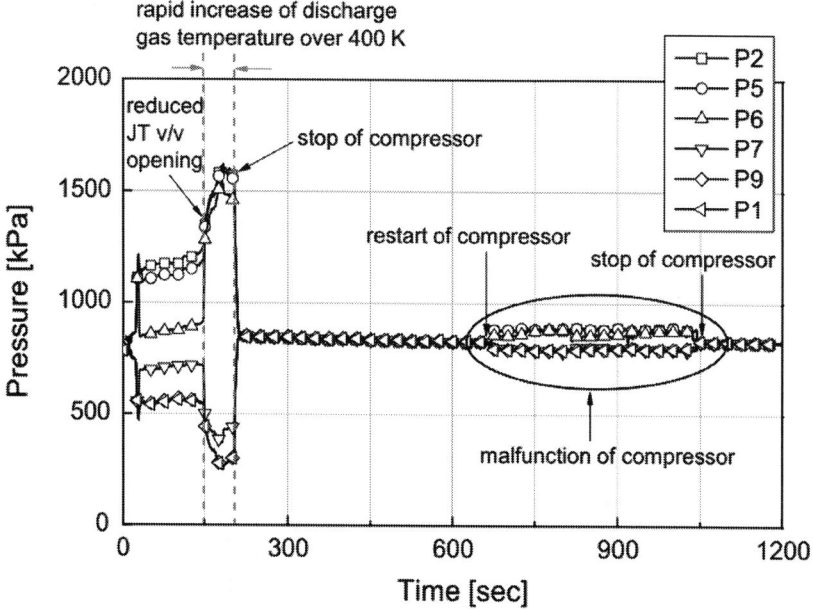

Figure 4: Pressure behavior of MR J–T operation without by-pass line.

The cool down characteristics of the MR J–T cryocooler are presented in Fig. 5. The compression ratio increased up to 7.4 after 100 s of compressor operation, and this undoubtedly resulted in compressor overheating. Thus, the by-pass valve was regulated to allow a portion of the high-pressure stream to flow directly into the compressor suction port. In particular, by-pass valve openings were controlled during the early stage of the operation such that the compressor discharge temperature remained below 385 K. After the compressor overheating phenomenon had been overcome, the by-pass circuit was completely closed, and stable operation was achieved.

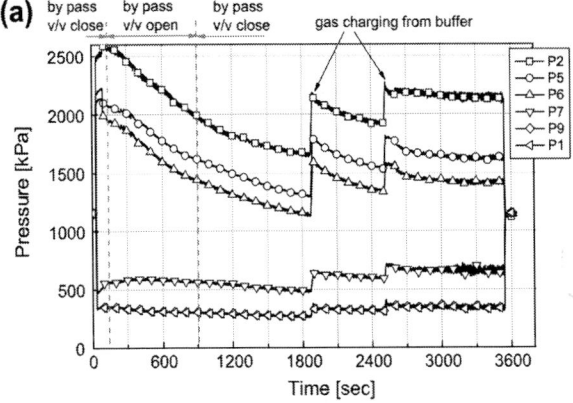

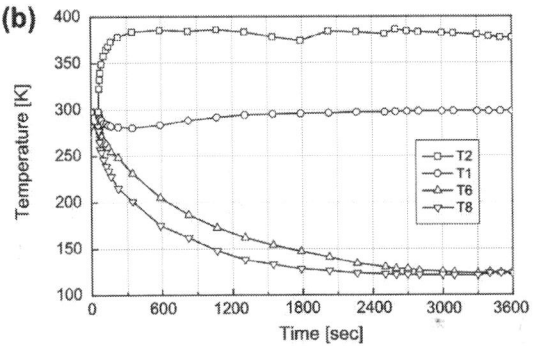

Figure 5: Cool down characteristic of MR J–T cryocooler with by-pass line; (a) pressure and (b) temperature.

Overall refrigerant pressures in the high pressure stream were found to gradually decrease as refrigerant density at low temperature decreased (Fig. 5a). Thus, after approximately 30 min of compressor operation, the mixed refrigerant gas was supplemented from the buffer reservoir. As a result, compressor discharge and suction pressures were slightly increased. The refrigeration system reached a steady state at around 55 min of compressor operation. The no load minimum temperature was 121 K, and the minimum temperature was 123 K with 5 W of thermal load. Typical pressures and temperatures during steady state operation are summarized in Table 1.

Table 1: Pressure and temperature at steady state operation

Stream	2	5	6	7	8	9
Pressure (kPa)	2150	1620	1440	660	–	350
Temperature (K)	373	298	123	–	123	293

The analyzed composition of the sampled MR by gas chromatography was 5.67, 30.71, 39.84, 9.26, and 14.52 mol% for nitrogen, methane, ethane, propane, and iso-butane (Table 2). Ethane was found to be present at more than the design level, whereas nitrogen composition was slightly lower than designed.

Table 2: Comparison of designed MR composition and analyzed composition of sampled MR by GC

Component	Nitrogen	Methane	Ethane	Propane	Iso-butane
Standard gas 1 (mol %)	11.96	34	34.02	8.01	12.01
Standard gas 2 (mol %)	17.95	26	26.03	12.01	18.01
Designed MR composition (mol %)	15	30	30	10	15
Analyzed MR composition (mol %)	5.67	30.71	39.84	9.26	14.52

DISCUSSION

Based on a series of preliminary experiments, we found that compressor overheating was potentially a serious problem when the MR J–T was operated using a commercial scroll compressor. Undoubtedly the authors paid special attention to the discharge temperatures, particularly during the early stage of experiments. However, as expected, the temperature soared steeply over 400 K with the compressor running, and it leads to the burn-out of the tip seals and irreversible malfunction of the scroll compressor.Fig. 6 describe the tip seal of the orbiting scroll that was burned out because the compressor discharge temperature was overshot. Compression using scroll compressors is achieved using two spiral scrolls. However, the scroll tip seal was found to be most vulnerable to overheating that the maximum allowable discharge gas temperature is limited by 394 K [10]. This overheating is inherently affected by the specific heat ratio ($k = C_p/C_v$) of the refrigerant. For ideal gases, gas temperature is proportional to compression ratio during isentropic compression, as expressed in the following equation:

$$\frac{T_2}{T_1} = \left(\frac{P_2}{P_1}\right)^{(k-1)/k}$$

(1)

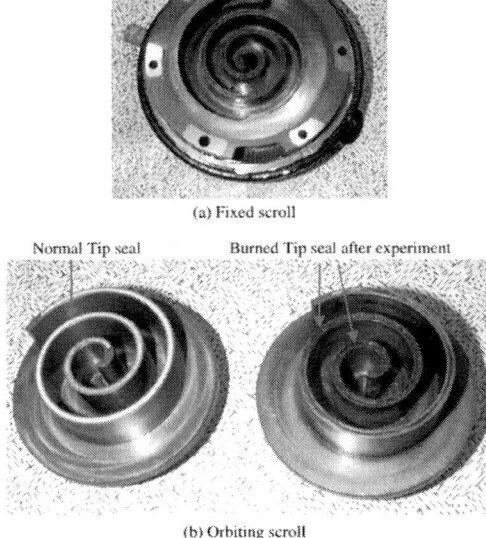

(a) Fixed scroll

Normal Tip seal   Burned Tip seal after experiment

(b) Orbiting scroll

Figure 6: Configuration of (a) fixed scroll, and (b) comparison of normal tip seal and burned tip seal of orbiting scroll.

As listed in Table 3[11], the specific heat ratios of nitrogen and methane are higher than those of ethane, propane, and iso-butane, and even of R22. Therefore, it can be presumed that compression heat is mainly generated by nitrogen and methane. To solve this problem, an auxiliary water injection cooling unit (Fig. 7) for the compressor surface and a by-pass line were employed. Compressor overheating problem was not controllable without the by-pass function even though the water injection cooling was applied through the entire experiment. It is evident that in this design the compressor dissipates its excess heat through the refrigerant by-pass into the suction port. In our studies, this allowed stable operation of the cryocooler without any significant damage. The by-pass line was also found to be beneficial in terms of controlling the compression ratio during the early running period when the J–T expansion orifice opening is fixed.

Table 3: Specific heat ratio at 100 kPa and 300 K [11]

Refrigerant	Nitrogen	Methane	Ethane	Propane	Iso-butane	R22
Specific heat ratio k=Cp/Cv	1.401	1.305	1.193	1.135	1.104	1.184

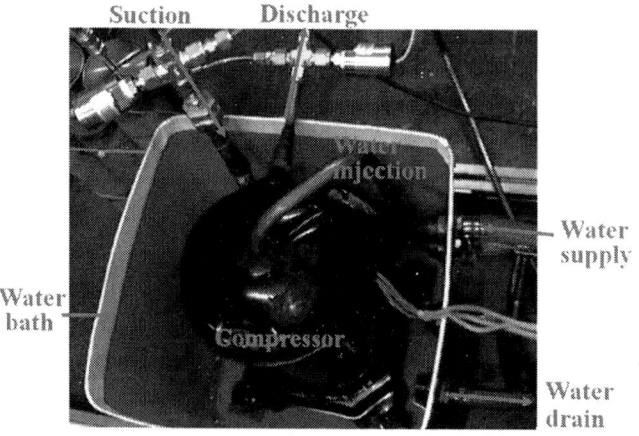

Figure 7: Configuration of water injection cooling unit for compressor surface.

A special care should be exercised with respect to oil management in cryocooling systems. This is primarily because oil particles in the refrigerant can freeze at low temperatures and eventually cause refrigerant clogging. For example, when we performed preliminary J–T experiments without oil mist filters, an unexpected pressure increase because of clogging was observed. During preliminary experiments, we had tried to install the oil separator prior to or after the aftercooler, and it was found that oil was effectively removed when the oil separator was placed upstream of the aftercooler. This would be particularly true because the fluidity of the lubrication oil could be improved with decreased viscosity at higher temperature. Furthermore, refrigerants, e.g., iso-butane, could be condensed during after cooling, which results in the reduction in oil separation efficiency. Therefore, the authors decided to install the oil separator prior to the aftercooler and the oil separation was found to be satisfactory throughout without any procedural malfunction. In addition to commercial oil separators, we integrated domestically manufactured oil mist filters in the circuit. In particular, the oil mist filter includes three stages of glass microfiber filters to ensure satisfactory oil mist removal while allowing a moderate pressure drop through the filters.

Despite the use of a novel by-pass circuit to ensure stable operation of the MR J–T cryocooler, further optimization of the MR J–T unit is required. This is principally because of the difference between the lowest temperature achieved (121 K) in the present study and the calculated value (94 K). This discrepancy presumably results from the large pressure loss through the high-pressure stream, due to the additional oil mist filters integrated in the circuit and the heat exchanger. The pressure drop through the oil separator and oil mist filters was approximately 530 kPa. In addition, the pressure drop through the heat exchanger was 180 and 260 kPa for high-pressure and low-pressure streams, respectively. Consequently, the expanded pressure at the evaporator was maintained at 660 kPa, which caused a higher minimum temperature than expected.

In addition, the difference in minimum temperature of the cryocooler between the design and experiment value could be ascribed to different mixture compositions. As described previously, ethane was found to be present at a level more than that required, whereas the nitrogen level was lower. Each component of the mixed refrigerant was charged into the system using pressure ratios corresponding to mole fractions based on an ideal gas assumption. Based on the normal boiling points of components, the iso-butane was first filled into the system followed by propane, ethane, methane, and nitrogen, until the total circuit pressure reached the desired level. However, during the charging procedure, a steady state pressure was hardly achieved with fluctuation especially for high boiling point components. Considering this effect, it took more than 20 h to charge iso-butane and propane. However, ethane, methane, and nitrogen were charged into the system relatively quickly because they reached at steady pressure level quickly. It was assumed that the ethane exceeded the desired value (Table 2) due to its short charging time. As shown in Fig. 8, the effects of pressure loss and composition differences on J–T cryocooler performance are not trivial. The enthalpies of sampled refrigerant compositions were calculated at measured pressures of 1440 and 660 kPa, and enthalpy differences were plotted in temperature. The graph was found to be shifted right and toward the temperature axis as compared with the graph in Fig. 3, and the cooling capacity approached zero at 120 K, which corresponds to the lowest acquired temperature (121 K) in the present study.

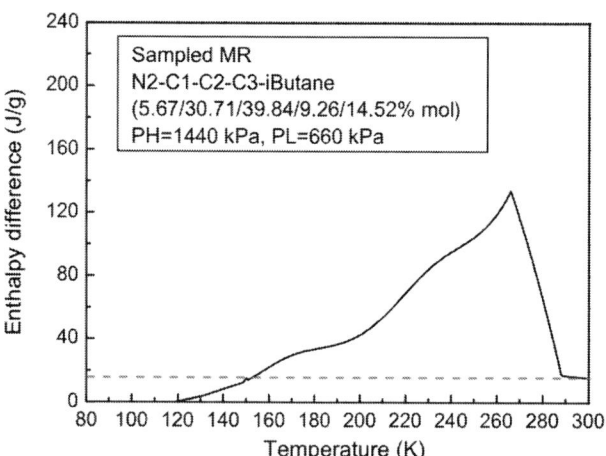

Figure 8: Enthalpy difference between high pressure and low pressure with sampled mixture [11].

The fundamental function of the buffer reservoir is to supplement refrigerant into the refrigeration circuit. The initial temperature of the

J–T cryocooler system was ambient. However, the heat exchanger, J–T expansion, and evaporator are gradually cooled during operation, and some portion of refrigerant is condensed and liquefied, which decreases total system pressure and cooling capacity. Accordingly, to increase the cooling effect of the MR J–T system, the buffer reservoir could be used, and this was found to an effective means of providing additional cooling capacity, especially when the compressor discharge pressure was decreased.

Further studies are required to optimize the MR J–T cycles obtained using a scroll compressor with various refrigerants. In fact, the optimization of compression characteristics including neon could be more challenging because the specific heat ratio of neon is higher than that of hydrocarbons. However, neon–nitrogen mixtures would be more appropriate for a lower temperature range J–T cryocooler. Although, we do not discuss about the refrigeration efficiency of the MR J–T cryocooler, an investigation of compressor characteristics and efficiency regarding to the refrigeration cycle coefficient of performance (COP) is necessary. In addition, significant pressure drop mainly through the oil mist filters and the heat exchanger should be improved to optimize the cycle efficiency. The large pressure drop particularly through the oil mist filters could be improved using commercial oil mist filters which are highly optimized against the pressure drop. Furthermore, it is also true that the pressure drop through the heat exchanger could be markedly reduced by using tubes with larger diameters.

CONCLUSIONS

The present study demonstrates the feasibility of a MR J–T cryocooler operating with a commercial oil lubricated scroll compressor. By providing the MR J–T system with novel features, such as, a by-pass path, especially during the early stage of operation, permanent damage to the compressor due to the overheating can be avoided and stable operation ensured. In addition, a new oil filtering system utilizing glass microfiber filters was devised to prevent the passage of oil particles into the expansion unit of the system. Therefore, we are of the opinion that commercial oil lubricated scroll compressors can be used for MR J–T cryocoolers, but caution that the compressor overheating phenomenon must be addressed.

ACKNOWLEDGEMENTS

This work was supported by the Power Generation & Electricity Delivery of the Korea Institute of Energy Technology Evaluation and Planning (KETEP) grant funded by the Korea government Ministry of Knowledge Economy (No. 2011101050002B).

REFERENCES

1. Radebaugh R. Cryocoolers: the state of the art and recent developments. J Phys: Condens Matter 2009; 21:164219.
2. Walimbe N, Narayankhedkar K, Atrey M. Experimental investigation on mixed refrigerant Joule–Thomson cryocooler with flammable and non-flammable refrigerant mixtures. Cryogenics 2010; 50:653–9.
3. Lee J, Hwang G, Jeong S, Park BJ, Han YH. Design of high efficiency mixed refrigerant Joule–Thomson refrigerator for cooling HTS cable. Cryogenics 2011; 51:408–14.
4. Lin MH, Bradley PE, Marcia LH, Lewis R, Radebaugh R, Lee YC Mixed refrigerants for a glass capillary micro cryogenic cooler Cryogenics 2010; 50:439–42.
5. Boiarski MJ, Brodianski VM, Longsworth RC. Retrospective of mixed-refrigerant technology and modern status of cryocoolers based on one-stage, oillubricated compressors. Adv Cryog Eng 1998; 43:1701–8.
6. Luo EC, Zhou Y, Gong MQ Mixed-refrigerant Joule–Thomson cryocooler driven by R22/R12 compressor Adv Cryog Eng 1998; 43:1675–8.
7. Gong M, Wu J, Luo E, Qi Y, Zhou Y. Study of the single-stage mixed-gases refrigeration cycle for cooling temperature-distributed heat loads. Int J Therm Sci 2004; 43:31–41.
8. Reddy KR, Murthy SS, Venkatarathnam G. Relationship between the cooldown characteristics of J–T refrigerators and mixture composition. Cryogenics 2010; 50:421–5.
9. Radebaugh R. Recent developments in cryocoolers in: 19th International congress of refrigeration; 1995. p. 973–89.

10. Emerson Climate Technologies, European Headquarters, Pascalstrasse 65, 52076 Aachen, Germany. <http://www.emersonclimate.com/en-us/resources/ faq/Pages/faq.aspx>.

11. Lemmon EW, Huber ML, McLinden MO. NIST standard reference database 23: reference fluid thermodynamic and transport properties-REFPROP, Version 9.0. Gaithersburg: National Institute of Standards and Technology, Standard Reference Data Program; 2010.

Citations

CHAPTER 1

Hongkun Li, Xuefeng Zhang, Xiaowen Zhang, Shuhua Yang, and Fujian Xu, "Pressure Pulsation Signal Analysis for Centrifugal Compressor Blade Crack Determination," Mathematical Problems in Engineering, vol. 2014, Article ID 862065, 15 pages, 2014. doi:10.1155/2014/862065.

CHAPTER 2

Henning Friege, The role of waste management in the control of hazardous substances: lessons learned, doi:10.1186/2190-4715-24-35.

CHAPTER 3

Farhat, S. , Ouar, N. , Hosni, M. , Hinkov, I. , Mercone, S. , Schoenstein, F. and Jouini, N. (2014) Scale-Up of the Polyol Process for Nanomaterial Synthesis. Journal of Materials Science and Chemical Engineering, 2, 1-11. doi: 10.4236/msce.2014.29001.

CHAPTER 4

J. Sołoducho, K. Olech, A. Świst, D. Zając and J. Cabaj, «Recent Advances of Modern Protocol for C-C Bonds—the Suzuki Cross-Coupling,» Advances in Chemical Engineering and Science, Vol. 3 No. 3A, 2013, pp. 19-32. doi: 10.4236/aces.2013.33A1003.

CHAPTER 5

J. Khorshidi and H. Davari, "Experimental Study of Drying Process of COLZA Seeds in Fluidized Bed Dryer by Statistical Methods," Advances in Chemical Engineering and Science, Vol. 2 No. 1, 2012, pp. 129-135. doi:10.4236/aces.2012.21016.

CHAPTER 6

Rabelo, M., Bellato, C., Silva, C., Ruy, R., Silva, C. and Nunes, W. (2014) Application of Photo-Fenton Process for the Treatment of Kraft Pulp Mill Effluent. Advances in Chemical Engineering and Science, 4, 483-490. doi:10.4236/aces.2014.44050.

CHAPTER 7

Seung Min Yeo, Andreas A. Polycarpou, Tribological performance of PTFE- and PEEK-based coatings under oil-less compressor conditions, Wear, Volume 296, Issues 1–2, 30 August 2012, Pages 638-647, ISSN 0043-1648, http://dx.doi.org/10.1016/j.wear.2012.07.024.

CHAPTER 8

Fritz Klocke, Andreas Klink, Drazen Veselovac, David Keith Aspinwall, Sein Leung Soo, Michael Schmidt, Johannes Schilp, Gideon Levy, Jean-Pierre Kruth, Turbomachinery component manufacture by application of electrochemical, electro-physical and photonic processes, CIRP Annals - Manufacturing Technology, Volume 63, Issue 2, 2014, Pages 703-726, ISSN 0007-8506, Doi.org/10.1016/j.cirp.2014.05.004.

CHAPTER 9

Jisung Lee, Kyungsoo Lee, Sangkwon Jeong, Experimental study of a mixed refrigerant Joule–Thomson cryocooler using a commercial air-conditioning scroll compressor, Cryogenics, Volumes 55–56, May–July 2013, Pages 47-52, ISSN 0011-2275, http://dx.doi.org/10.1016/j.cryogenics.2013.02.001.

Index